# Engine Service

W. GARY LEWIS, 1941-

PRENTICE-HALL, INC., Englewood Cliffs, New Jersey 07632

*Library of Congress Cataloging in Publication Data*

LEWIS, W GARY

Engine service.

Includes bibliographical references.

1. Automobiles–Motors–Maintenance and repair. I. Title.

TL210.L47 629.2′504′028 79-27920

ISBN 0-13-277236-1



Interior design and production by M. L. McAbee
Cover design by Prudence Kohler
Manufacturing buyer: Gordon Osbourne

Printed in the United States of America

10 9 8 7 6 5 4 3 2 1

Prentice-Hall International, Inc., *London*
Prentice-Hall of Australia Pty. Limited, *Sydney*
Prentice-Hall of Canada, Ltd., *Toronto*
Prentice-Hall of India Private Limited, *New Delhi*
Prentice-Hall of Japan, Inc., *Tokyo*
Prentice-Hall of Southeast Asia Pte. Ltd., *Singapore*
Whitehall Books Limited, *Wellington, New Zealand*

# Contents

# Preface

This text has been developed to provide student mechanics and apprentices with "state of the art" information on engine overhauling and rebuilding. It shows how to perform engine overhauls and engine rebuilding and explains why particular techniques are followed.

Much of this information is written from the viewpoint of the automotive machinist serving automotive mechanics. The machine shop view of service procedures is particularly important, because so many mechanics from dealerships and independent garages now sublet their machining operations to the automotive machine shops.

The procedures that have been included have been selected carefully from those with wide industry acceptance and represent a well-balanced current and competent "state of the art" practice. It has been my experience, both in performing and teaching engine service, that the procedures discussed here are among the most satisfactory, both to instructor and student.

While attention is given to theory and to explanations of why certain procedures are followed, the emphasis throughout is upon "how to" information. This text should certainly help the student-apprentice to better comprehend and use the automotive service manuals that have been written by the manufacturers for experienced mechanics. After reading this text, the student apprentice should be much better attuned to the important details of servicing, which are easily overlooked.

This text helps the instructor fill the gap between the more theoretical automotive texts and the purely technical service manuals, enabling the instructor to guide successive classes of students into making better judgments. This shop work can proceed more rapidly and with fewer oversights because student apprentices are better able to select a correct method of repair and be more careful in carrying it out.

W. GARY LEWIS

# Acknowledgements

The author expresses sincere gratitude to the following companies and corporations: General Motors Corporation, Chrysler Corporation, Sioux Tools Incorporated, Kwik-Way Manufacturing Company, Sunnen Products Company, Tobin-Arp Manufacturing Company, Van Norman Machine Company, Bear Manufacturing Company, K. O. Lee Company, Fel-Pro Incorporated, Bowman Distribution of the Barnes Group Incorporated, Brown Sharpe Manufacturing Company, and Stewart Warner Corporation.

Sincere gratitude is also given to the following individuals: Frank Berger of Forberg Automotive; Don Arthur of Penniman and Richards Machine Shop; Chico Guthrie of Cal-Cams; Bud Riebhoff, automotive machinist instructor, De Anza College; and Tim Mitchell, graphics artist, De Anza College.

# 1

# Engine Theory

The information covered under "Engine Theory" is intended to support the technical content of this text. Theory will be stated briefly and will be limited to essential points. Comprehension of this information will aid in the diagnosis of engine malfunctions and in making critical judgments regarding engine service.

## THE FOUR-STROKE CYCLE

Automotive engines, with few exceptions, operate on a four-stroke cycle. An air-fuel mixture is drawn into a cylinder, compressed, and ignited. Upon ignition, gases expand and force the piston downward in the cylinder. Force and motion are transmitted from the piston through the connecting rod to the crankshaft. In this manner, reciprocating, or up and down, motion at the piston is changed to rotary motion at the crankshaft. The relationship of valves, piston, and crankshaft are as follows:

1. On the intake stroke, the intake valve is open, and the piston travels downward. The air-fuel mixture is forced into the cylinder because of low pressure in the cylinder and higher atmospheric pressure outside the engine (see Fig. 1-1).

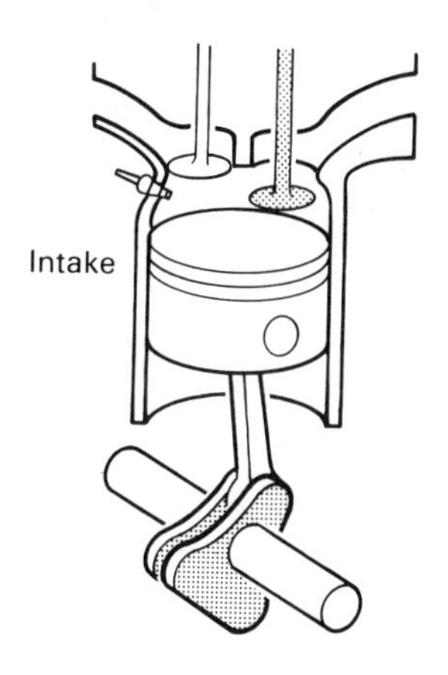


*FIGURE 1-1 The intake stroke*

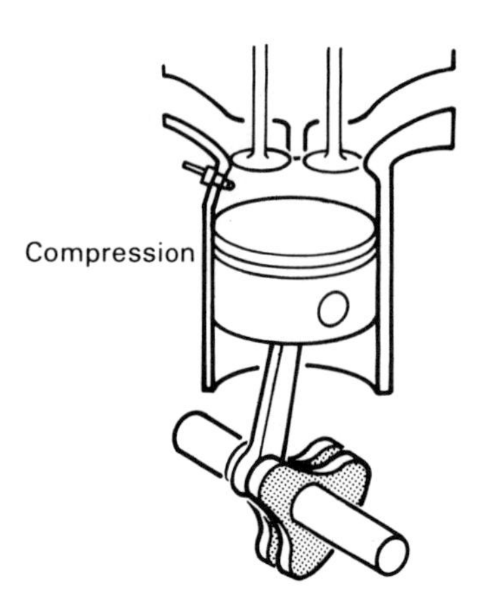


*FIGURE 1-2 The compression stroke*

2. On the compression stroke, both intake and exhaust valves are closed, and the piston travels upward in the cylinder. The piston travel compresses the air-fuel mixture; then ignition occurs (see Fig. 1-2).

3. On the power stroke, both intake and exhaust valves remain closed. Upon ignition of the air-fuel mixture, the expansion of burning gases forces the piston to travel downward in the cylinder (see Fig. 1-3).

4. On the exhaust stroke, the exhaust valve is open, and the piston travels upward in the cylinder. Burned gases are forced through the exhaust valve by the piston (see Fig. 1-4).

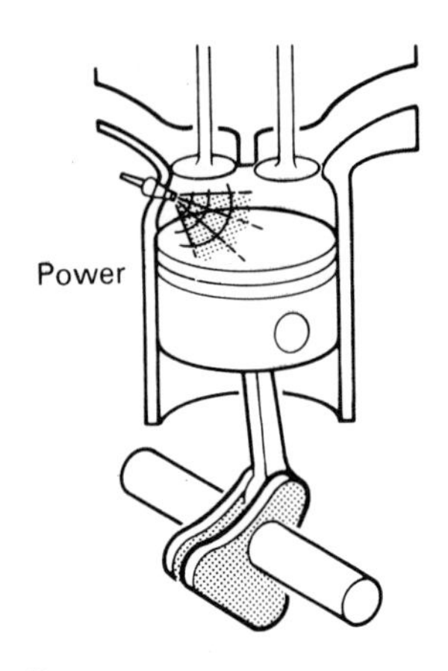


*FIGURE 1-3 The power stroke*

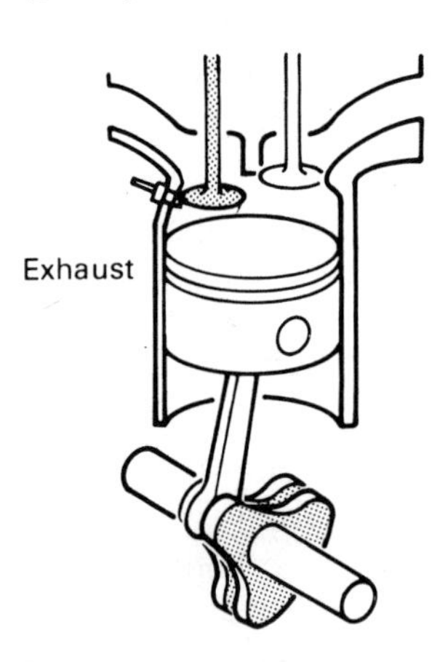


*FIGURE 1-4 The exhaust stroke*

Keep in mind that one stroke requires one-half turn, or 180° of crankshaft rotation. Four strokes requires two full turns, or 720° of crankshaft rotation. All cylinders, regardless of the number of cylinders in the engine, complete the four-stroke cycle in two crankshaft revolutions.

## VALVE TIMING AND CAMSHAFTS

Because each valve must open and close every two crankshaft revolutions, the camshaft is driven at one-half crankshaft speed. Therefore, camshaft drive sprockets or drive gears have twice as many teeth as crankshaft sprockets or gears (see Figs. 1-5 and 1-6).

In the description of the four-stroke cycle just given, periods of valve opening or valve closing are not detailed. In fact, intake and exhaust valves are each open for more than one stroke as measured at the crankshaft so that high and low pressure conditions can be used to promote gas flow in and out of the cylinder. For example, the intake valve is open for approximately 260° of crankshaft rotation because cylinder pressure remains low relative to atmospheric pressure during this period. The exhaust valve is also open for approximately 260° (see Fig. 1-7). The period when each valve is open is called *duration.*

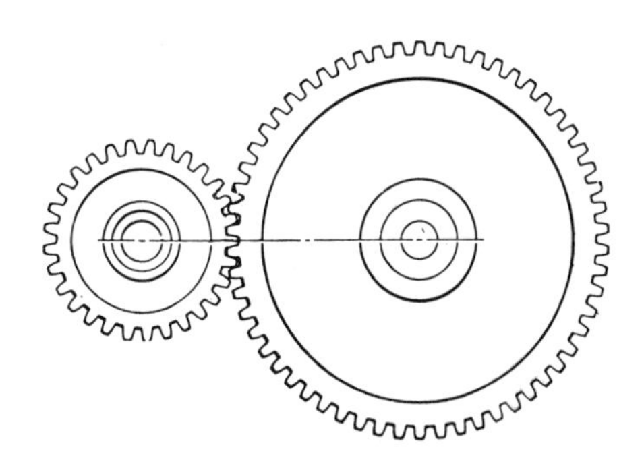

*FIGURE 1-5 Gear-driven camshaft*

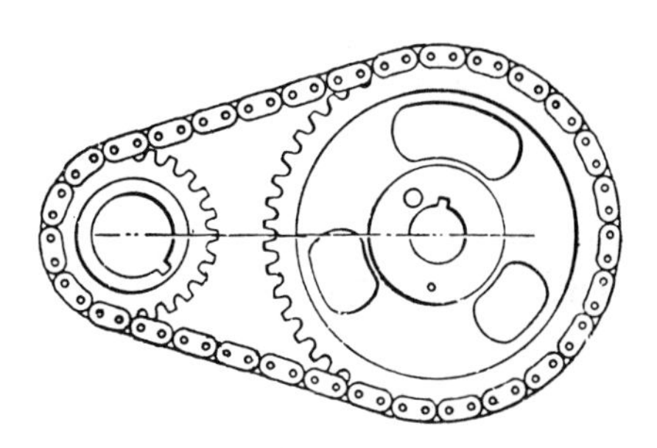

*FIGURE 1-6 Chain-driven camshaft*

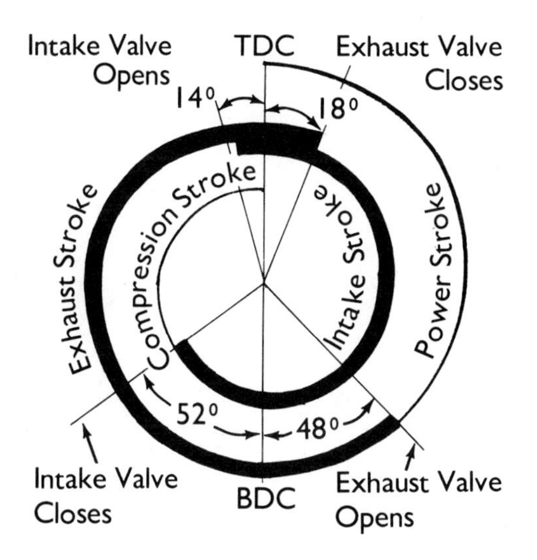


*FIGURE 1-7 Valve timing diagram*

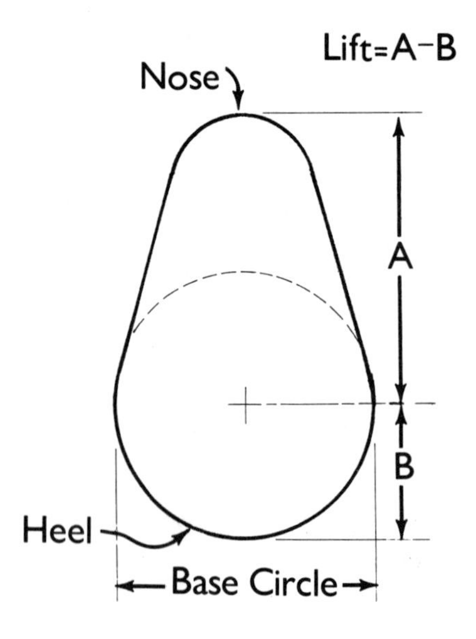


*FIGURE 1-8 Cam lift*

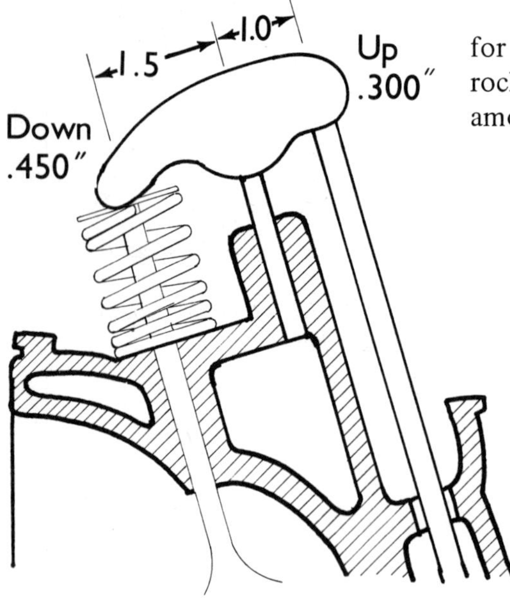


*FIGURE 1-9 Rocker arm ratio*

It can be seen in Figure 1-7 that there is a period when both intake and exhaust valves are open at the same time. This period is called *valve overlap* and occurs at the end of the exhaust stroke and the start of the intake stroke. Valve overlap promotes gas flow because lower pressure at the exhaust port draws air and fuel, at higher pressure, through the intake port. As the exhaust valve closes, the piston continues downward, creating a lower pressure in the cylinder that continues to draw in air and fuel.

Another common term relating to valve systems is *cam lift*. Cam lift is the total range of travel for a valve lifter from the valve closed to the valve open position. This is determined by calculating the difference between the distance from the nose of the cam to the center of rotation and the distance from the heel of the cam, located on the *base circle*, to the center of rotation (see Fig. 1-8).

*Valve lift* is usually greater than cam lift for engines using rocker arms. This is because rocker arms can act as levers to open valves an amount greater than the cam lift (see Fig. 1-9).

The rocker arm ratio, the ratio between valve lift and cam lift, is commonly between 1.5:1 and 1.75:1.

## VALVE LIFTERS

There are two common types of valve lifters in use today. *Solid valve lifters* have no internal parts and require clearance, or *lash,* in the valve train mechanism to ensure closing of the valve (see Fig. 1-10). *Hydraulic valve lifters* are designed to maintain *zero lash* in the valve train mechanism. The advantages of hydraulic lifters include quieter engine operation and the elimination of the periodic adjustment required to maintain proper clearance for solid valve lifters. Hydraulic valve lifters do, however, maintain a constant pressure on the camshaft that solid valve lifters do not; therefore, the antiscuff properties of lubricating oils are more critical with hydraulic lifters.

Hydraulic valve lifters maintain zero lash as follows:

1. In the valve closed position, oil flows through the lifter body to the interior of the plunger and past the check valve (see Fig. 1-11). Zero lash is maintained by the plunger spring at this point.
2. As the cam raises the lifter body to open the valve, the check valve seats, and trapped oil limits slippage of the plunger within the lifter body (see Fig. 1-12). The valve mechanism is opened just as it would be with a solid valve lifter.
3. As the camshaft returns to the valve closed position, oil is again fed through the lifter body, through the plunger, and past the open check valve. This is how oil lost by leakage between the plunger and lifter

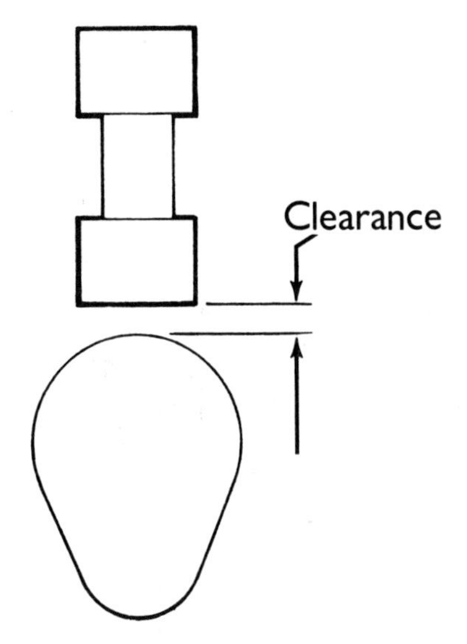


*FIGURE 1-10 Valve lash, or clearance, with solid valve lifters*

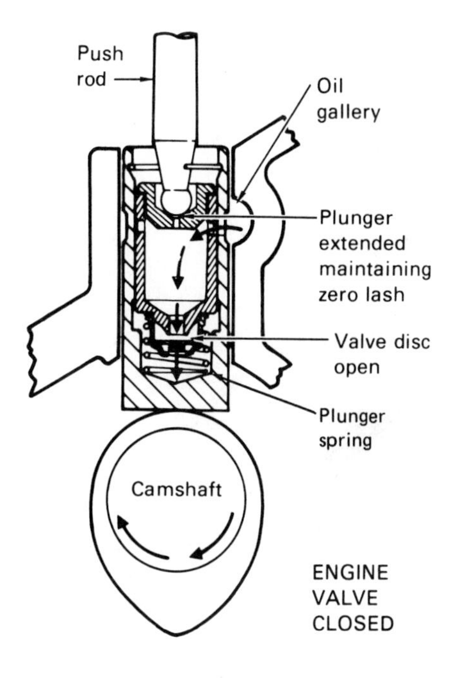


*FIGURE 1-11 Oil flow through a hydraulic lifter in the valve closed position*

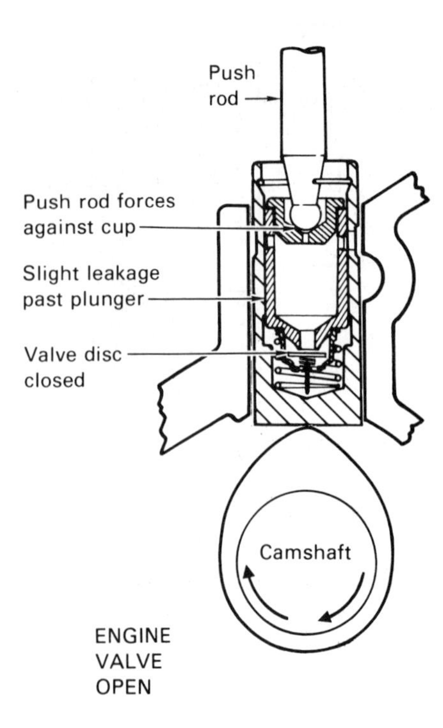


*FIGURE 1-12 Oil flow through a valve lifter in the valve open position*

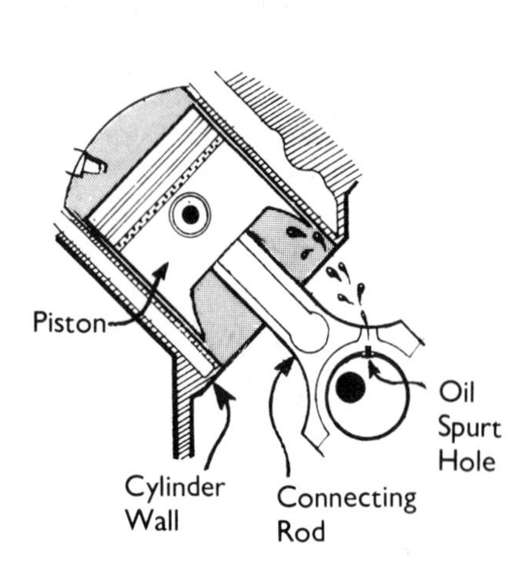


*FIGURE 1-13 Oiling cylinder walls with spray*

body is replaced. This leakage is called *predetermined leakdown.*

## ENGINE OILING

Lubrication prevents metal-to-metal contact between moving parts. Lubrication is provided by combinations of spray and pressurized oil circulation (see Figs. 1-13 and 1-14).

Engine oil also acts to cool engine parts. Oil cools pistons and cylinders as it drains back into the oil pan. The pressurized oiling of engine bearings also aids in cooling as the oil absorbs heat from the crankshaft and bearing surfaces.

Oil is stored in the *oil sump.* The sump is the lowest point in the oil pan. Oil pans are often fitted with *baffles* to keep oil in the sump on hard braking, acceleration, or cornering (see Fig. 1-15). A *windage tray* may also be positioned between the crankshaft and the oil sump to prevent turbulent air flow from

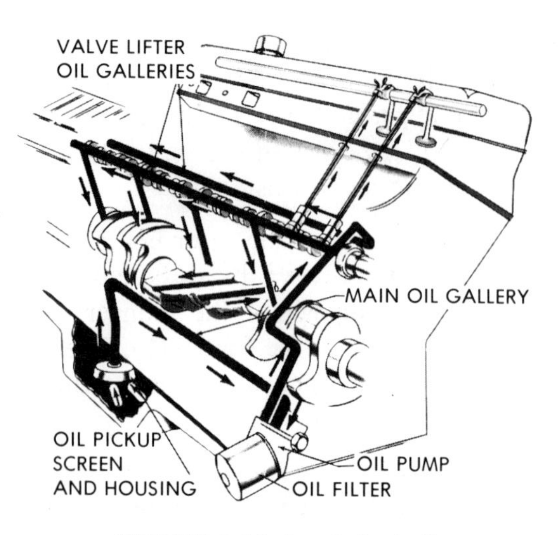


*FIGURE 1-14 Pressurized oil circulation (Courtesy of Buick Motor Div.)*

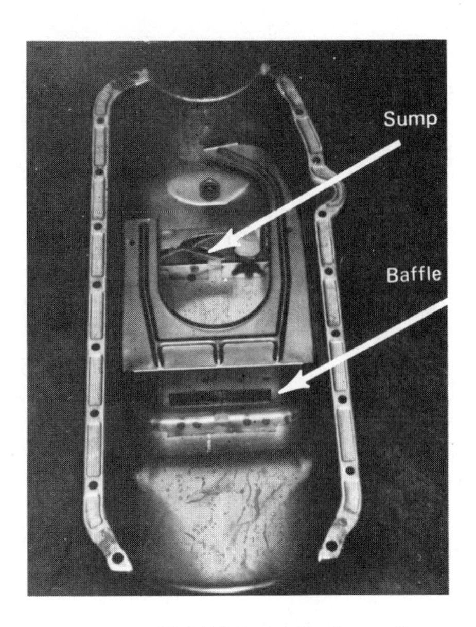


*FIGURE 1-15 An oil pan with baffles*

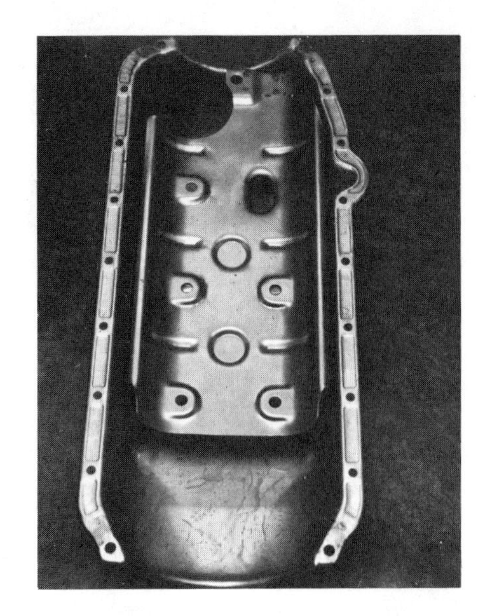

*FIGURE 1-16 An oil pan and windage tray*

the rotating crankshaft from affecting oil in the sump (see Fig. 1-16).

Oil is circulated by a pump mounted in the engine crankcase or possibly in the front timing cover or on the outside of the block. The pump draws oil from the sump of the oil pan through a tube extending from the pump inlet to the sump. The sump end of the pickup tube is fitted with a screen to prevent larger pieces of sediment from entering the pump (see Fig. 1-17).

*FIGURE 1-17 An oil pump (in the crankcase) and pickup tube (Courtesy of Chrysler Corp.)*

*FIGURE 1-18 An oil pump driven by the distributor shaft (Courtesy Pontiac Motor Div.)*

Most oil pumps in use now are driven by an extension of the distributor shaft (see Fig. 1-18) or by direct engagement of a pump drive gear with a gear on the camshaft (see Fig. 1-19).

Most oil pumps are gear pumps or rotor pumps. Gear pumps use a pair of meshing gears in a closed housing. One gear is driven by the pump drive and the second gear is driven by the first gear. A pressure relief valve is provided to regulate oil pressure (see Fig. 1-20). In rotor pumps, the inner rotor is driven by the oil pump drive and an outer rotor is driven by the inner rotor (see Fig. 1-21). As with the gear pumps, the oil pressure at the outlet is regulated by a pressure relief valve.

*FIGURE 1-19 An oil pump driven by a gear of the camshaft (Courtesy of Chrysler Corp.)*

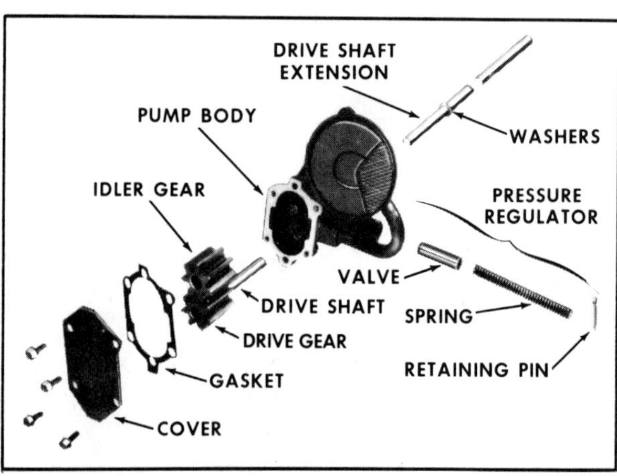


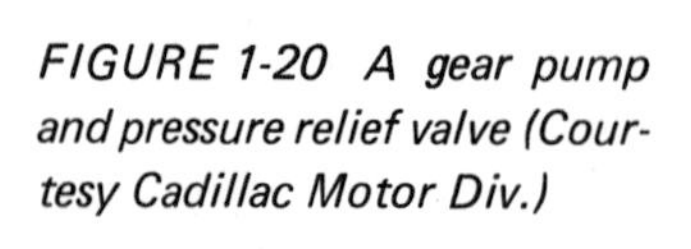

*FIGURE 1-20 A gear pump and pressure relief valve (Courtesy Cadillac Motor Div.)*

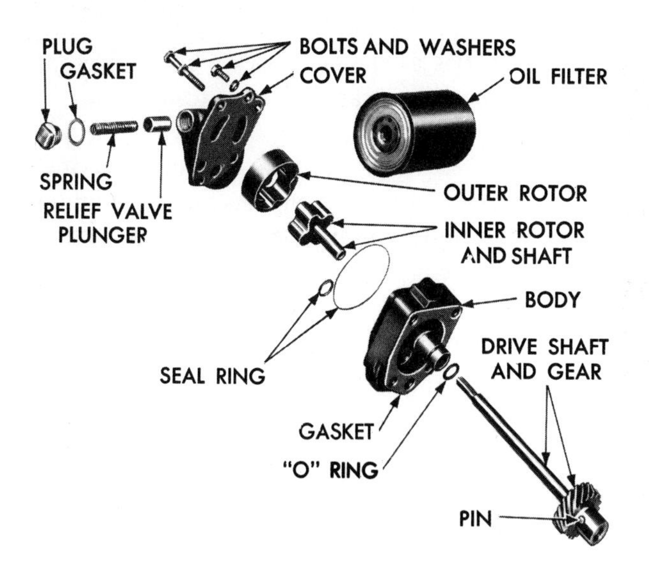


*FIGURE 1-21 A rotor pump and pressure relief valve (Courtesy Chrysler Corp.)*

Nearly all engines in use now use *full-flow oil filtering* systems. This means that all oil pump output goes through a filter before being circulated through the engine. Oil filters are typically made from a resin-treated paper or other materials with a very fine porosity. The resin treatment prevents contaminants in the oil, such as water or acids, from affecting the filter material. The porosity of the filter material must be fine enough to trap solid contaminants of any significant size but must at the same time be low enough in restriction so that oil flow is not inhibited. A bypass valve is incorporated into the filtering system so that a dirty and overly restrictive filter cannot limit engine oiling. The bypass valve opens when the filter is plugged, and oil pump output goes directly to the engine (see Fig. 1-22).

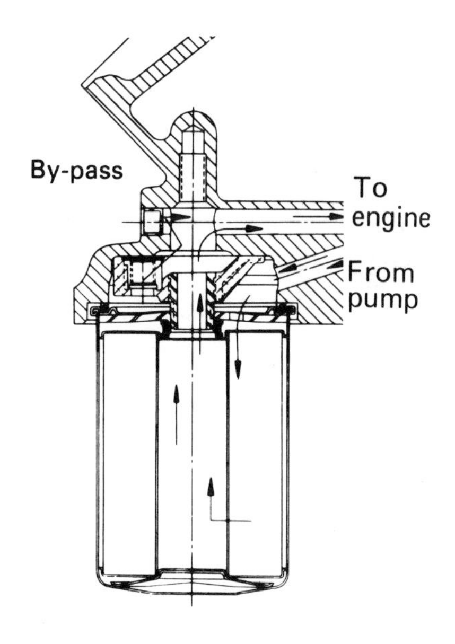


*FIGURE 1-22 An oil filter bypass circuit (Courtesy Buick Motor Division)*

## ENGINE OILS

The American Petroleum Institute, the Society of Automotive Engineers, and the American Society for Testing and Materials have established a series of service classifications for engine oils. New classifications may be added

to this series as the need arises, but at present the service classifications are as follows:

**SA** – Utility Gasoline Engine Service: Service typical of engines operated under such mild conditions that the protection afforded by compounded oils is not required. This classification has no performance requirements.

**SB** – Minimum Duty Gasoline Engine Service: Service typical of engines operated under such mild conditions that only minimum protection afforded by compounding is required. Oils designed for this service have been used since the 1930s and provide only antiscuff capability and resistance to oil oxidation and bearing corrosion.

**SC** – 1964 Gasoline Engine Warranty Service: Service typical of gasoline engines in 1964 through 1967 models of passenger cars and trucks operating under engine manufacturers' warranties in effect during those years. Oils designed for this service control high and low temperature deposits, wear, rust, and corrosion in gasoline engines.

**SD** – 1968 Gasoline Engine Warranty Service: Service typical of gasoline engines in passenger cars and trucks beginning with 1968 models and operating under engine manufacturers' warranties. Oils designed for this service provide more protection from high and low temperature engine deposits, wear, rust, and corrosion in gasoline engines than SC and may be used when SC is recommended.

**SE** – 1972 Gasoline Engine Warranty Service: Service typical of gasoline engines in passenger cars and some trucks beginning with 1972 and certain 1971 models operating under engine manufacturers' warranties. Oils designed for this service provide more protection against oil oxidation, high temperature engine deposits, rust, and corrosion in gasoline engines than SC and SD oils. SE oil may be used whenever SC or SD is recommended.

**CA** – Light-Duty Diesel Engine Service: Service typical of diesel engines operated in mild to moderate duty with high quality fuels. Occasionally has included gasoline engines in mild service. Oils designed for this service were widely used in the late 1940s and in the 1950s. These oils protect against bearing corrosion and high temperature deposits in normally aspirated diesel engines using fuels that do not impose unusual requirements for wear and deposit protection.

**CB** – Moderate-Duty Diesel Engine Service: Service typical of diesel engines operated in mild to moderate duty but with lower quality fuels that necessitate more protection from wear and deposits. Occasionally has included gasoline engines in mild service. Oils designed for this service were introduced in 1949. Such oils provide necessary protection from bearing corrosion and from high temperature deposits in normally aspirated diesel engines with sulfur fuels.

**CC** – Moderate-Duty Diesel and Gasoline Engine Service: Service typical of lightly supercharged diesel engines operated in moderate to severe duty and certain heavy duty gasoline engines. Oils designed for this service were introduced in 1961 and used in many trucks and in industrial, construction, and farm equipment. These oils protect against high temperature deposits in lightly supercharged diesels and also against rust, corrosion, and low temperature deposits in gasoline engines.

**CD** – Severe-Duty Diesel Engine Service: Service typical of supercharged diesel engines in high speed, high output duty requiring highly effective control of wear and deposits. Oils designed for this service were introduced in 1955 and provide protection from bearing corrosion and from high temperature deposits in supercharged diesel engines with fuels of wide quality range.

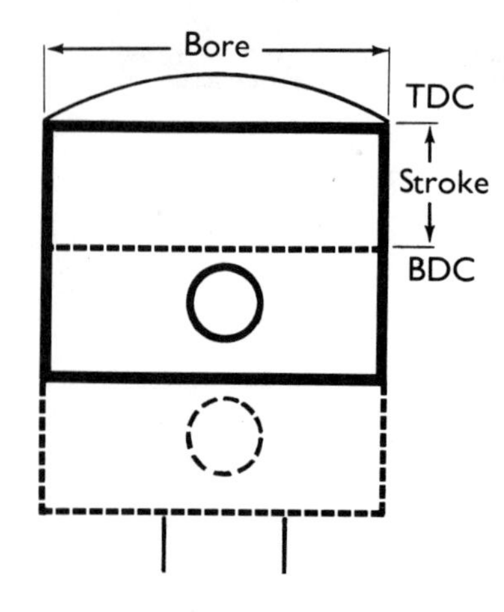


*FIGURE 1-23 Bore and stroke measurements*

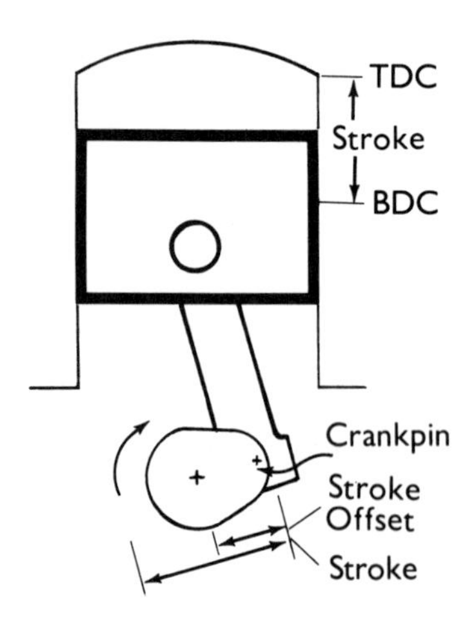


*FIGURE 1-24 Offset of the crankpin and stroke length*

## ENGINE MEASUREMENTS

The first engine measurements to become familiar with are *bore* and *stroke.* Bore is the diameter of the cylinder, and stroke is the distance between top dead center and bottom dead center of piston travel (see Fig. 1-23). When engine specifications are written, bore diameter is given first, and stroke length is given second.

The stroke *offset* of each crankpin of the crankshaft is one half of specified stroke length. Remember that a crankpin offset two inches from the center of rotation will cause the piston travel to be four inches per stroke (see Fig. 1-24).

*Displacement* is a calculation of the volume displaced by the piston travel in each cylinder. Total displacement is the sum of displacements for all cylinders in an engine. Displacement may be calculated for a single cylinder as follows:

$$\frac{\pi \times D^2 \times L}{4} = \text{volume}$$

$\pi = 3.1416$

D = diameter

L = stroke

*Compression ratios* are important because of their direct influence on engine efficiency. Compression ratios are calculated by dividing cylinder volume with piston at bottom dead center by cylinder volume with the piston at top dead center (see Fig. 1-25). It is important to note that for purposes of determining compression ratios, cylinder volumes include calculated displacement and any volume remaining above the piston at top dead center, including the combustion chamber volume of the cylinder head. This remaining volume is known as *clearance volume.*

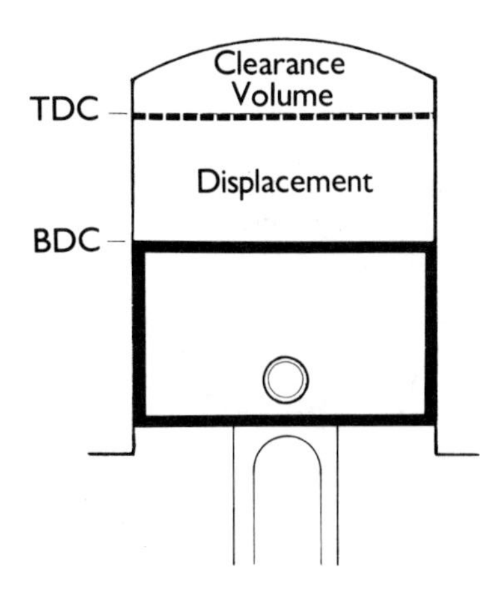


*FIGURE 1-25 Compression ratios are calculated by dividing the clearance volume into the displacement plus clearance volume.*

As mentioned, the compression ratio affects engine efficiency. Engines with higher compression ratios highly compress the air-fuel mixture before ignition occurs. This causes higher combustion pressures after ignition and increased power output and fuel economy. There is a trade-off in benefits, however, because higher octane fuels are required, and the higher combustion pressures also add to the stresses on piston rings and other components. Currently, lower compression ratios (below 9:1) are used to help reduce emissions of nitrogen oxides, which are associated with high combustion pressures and temperatures, and to make possible the use of unleaded fuel with reduced octane ratings.

## FITS AND CLEARANCES

The term *fit* has specific meaning and importance to engine mechanics and machinists. First, there are several types of fits used in any mechanism. There are *running* fits and *interference* fits to name just a couple. Each type of fit requires that sizes and dimensions of component parts be carefully measured to ensure optimum performance.

Crankshaft bearings are an example of running fits. *Clearance* is allowed between the

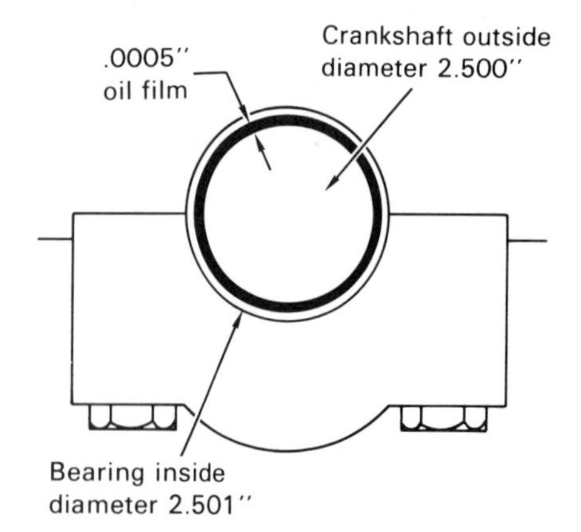


*FIGURE 1-26 The oil film on one side of a shaft with .002″ diametral clearance.*

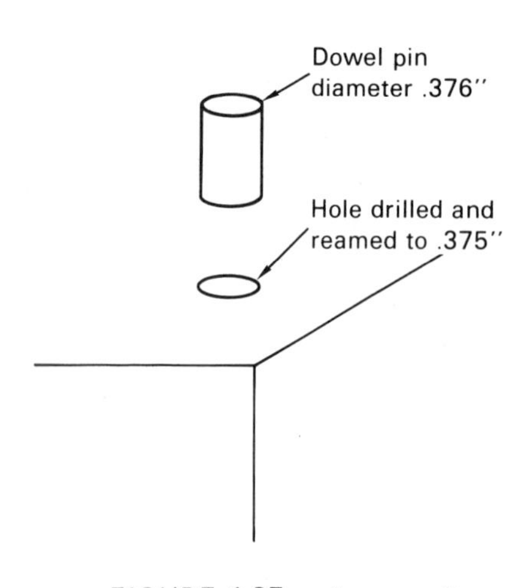


*FIGURE 1-27 A press-fit dowel pin.*

bearing surface and the shaft to permit oil flow. This clearance generally increases as the shaft diameter increases to permit a proportionate increase in oil flow through the bearing. With too little clearance, there will be metal-to-metal contact and wear. With too much clearance, there will be too much movement of the shaft, and knocking and bearing failure will result.

In other places, *interference,* or *press-fits* are used. Such fits are used to hold two parts in assembly. There is no relative motion between the two assembled parts. The outside diameter of the press-fit part is larger than the inside diameter of the part it is being forced into (see Fig. 1-26). For example, many piston pins are press-fit through the small end of the connecting rod.

Specified clearances are usually *diametral* clearances. This means that clearances are based on the diameter of the shaft as opposed to the radius. A specified .002-inch clearance (.05 mm) would mean a .001-inch clearance (.025 mm) on each side of the shaft (see Fig. 1-27).

To ensure that fits of either type fall within specifications, each part must be carefully measured. Each part has a *tolerance,* an acceptable variation in size. So long as parts are within tolerances, assemblies will be within specified clearance or interference limits unless selectively fitted.

*Selective fits* are frequently used to assemble parts, in which case, parts are matched to each other to obtain optimum fits. In this way, very precise assemblies can be made using more permissive levels of tolerance for individual parts. This practice reduces the cost of manufacturing for precision assemblies.

The major point of this part is the necessity to measure all engine parts so that fits are within specified limits. Precision measuring will do as much to maintain a high quality in engine service as any other single practice.

# 2

# Engine Diagnosis

Engine diagnosis must be as exact as possible so that satisfactory repairs can be made at minimal cost. Failure to perform complete tests and inspections generally leads to making repairs not required or failing to make required repairs, and both of these situations lead to dissatisfied customers.

One typical situation in which diagnosis is especially important is when valve grinding is being considered. This is because valve grinding will increase vacuum on the intake stroke. If rings are worn excessively, blow-by past piston rings will increase on compression strokes, and oil will bypass piston rings on intake strokes. While the engine may run smoothly, the customer will be unhappy with the increased oil consumption.

Avoid problems by testing engine condition before providing service. If tests of engine condition indicate poor ring sealing, valve grinding will be unsatisfactory without cylinder service. However, worn or heat-damaged oil seals (see Fig. 2-1) will account for considerable oil burning, especially through intake valve guides (see Fig. 2-2). If ring tests indicate good sealing, oil consumption rates will remain the same or even be reduced after valve service because of improved valve sealing.

A variety of tests for engine condition are given below. Of course, not all of these

*FIGURE 2-1 Valve stem seal hardened and cracked by heat*

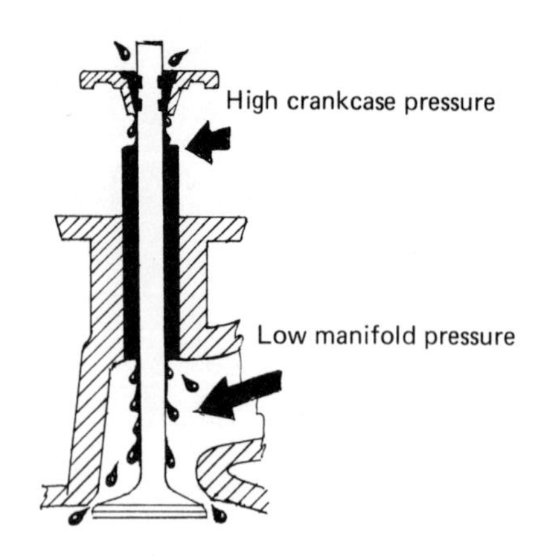


*FIGURE 2-2 Oil pull-over through an intake valve guide*

tests are going to be used on any one engine, but most repair facilities are equipped to perform at least some of these tests.

## OUTWARD SIGNS OF ENGINE WEAR

A worn engine gives outward signs of its condition. The most obvious of these is a blue-gray exhaust smoke that accompanies oil burning. However, the exhaust smoke associated with oil burning should not be confused with the black smoke caused by an overrich fuel mixture or the white condensation of water vapor in the exhaust. Oil burning caused by poor ring sealing will be most evident under acceleration, especially after the engine has been running at idle. Oil burning caused by oil passing through valve guides will be most evident on deceleration.

Mileage records of oil consumption should be kept and checked to help determine the extent of engine wear. Although normal rates of oil consumption vary widely among engines, a badly worn engine can be expected to use more than one quart of oil each 500 miles. A power loss also accompanies engine wear. The loss may go unnoticed by a "light-footed" driver, but it may be very noticeable to a more critical driver. Often the change is so gradual over a period of time that the power loss is not appreciated until the engine is reconditioned and producing full power again.

Check also for water in the oil or oil in the water. These conditions are proof of gasket failures or cracks in cylinder heads or blocks.

## BLOCK CHECKS

There is a test for combustion gases in the cooling system called a *block check.* The presence of gases can be caused only by cylinder head gasket leaks or cracks in engine castings. The block check test is performed by drawing vapors from the top of the radiator tank through a chemical test solution (see Fig. 2-3). The presence of combustion gases is confirmed by a change in solution color from blue to yellow.

*FIGURE 2-3 Testing for the presence of combustion gases in the cooling system*

## COMPRESSION TESTING

The compression test is probably the most widely used test of engine condition (see Fig. 2-4). Most repair shops will be equipped to perform compression tests, and testing can be completed within 30 minutes in most cases. The test basically measures the pressure produced in individual cylinders at cranking speed. From this, it can be determined whether the leakage is occurring past the piston rings or past the valves.

*FIGURE 2-4 Testing cylinder compression at cranking speed*

To separate piston ring leakage from valve leakage, the test is run "dry" and then "wet." If dry testing indicates proper compression pressures in each cylinder, further testing will not be required. If compression is low or varies beyond specifications, a wet test is in order. Compression is retested after injecting approximately a tablespoon of oil into each cylinder through the spark plug hole. If wet testing produces no changes in test results, the low compression is due to leaking valves. If wet testing raises test results by 10 percent or more, the compression loss is occurring past the piston rings.

Methods of interpreting compression tests vary widely. Some references will specify that compression tests should be within a 20 psi range; another will insist that test results should be within a 10 percent range. Some manufacturers specify that the low cylinder should be 75 percent or more of the high cylinder reading. It is recommended here that the specifications of each individual manufacturer be followed.

Compression testing is limited in that it cannot readily distinguish compression loss between exhaust or intake valves. Other tests, however, such as cylinder leakage or manifold vacuum testing, may be used to further isolate the source of leakage. As with other tests, compression testing may be misleading. For example, what may seem to be worn rings may turn out to be a scored cylinder wall. *Be sure to check the valve adjustment if testing indicates leaking valves.*

## CYLINDER LEAKAGE TESTING

*FIGURE 2-5 A cylinder leakage tester connected to an engine cutaway*

Cylinder leakage testing provides a more detailed analysis than compression testing. This test is performed by pressurizing each cylinder with compressed air, reading the percentage of total leakage on a gauge, and observing the source of escaping air (see Fig. 2-5).

For example, leaking piston rings are uncovered by listening for the escape of air from the crankcase oil-fill tube. A leaking exhaust valve is noted by escaping air at the exhaust pipe. A leaking intake valve is noted by escaping air at the carburetor. Worn valve guides are detected by blocking all crankcase openings and watching for air loss at the carburetor or exhaust pipe, depending on whether it is an intake or exhaust valve guide that has excessive clearance.

As with compression testing, determining what engine repairs are required is difficult. One reference, for example, specifies a maximum of 20 percent leakage. In practice, it is found that some engines have a 30 percent leakage rate and are still performing satisfactorily without noticeable power loss or oil consumption. Therefore, a relatively high rate of leakage can be considered acceptable and possibly normal for some engines. As with compression testing, check the valve adjustment if testing indicates leaking valves. It is recommended here that the 20 to 30 percent range be considered marginal. Check oil consumption rates and exhaust smoke before passing or condemning an engine that tests in the marginal zone.

## VALVE TIMING TESTS

Incorrect valve timing may be determined in several ways. First, the most common causes of incorrect valve timing should be considered. One is improper cam drive assembly, and another is the failure of sprockets (see Fig. 2-6), chains, or gears.

The obvious signs of incorrect valve timing include firing through the carburetor or exhaust system. Compression testing may result in low readings. However, it is possible for an engine with a timing chain that has "jumped a tooth" to run smoothly but with a drastic power loss.

Tests for valve timing generally involve checking the valve position relative to crankshaft position. For example, when the timing marks are at top dead center (TDC) on the compression stroke and the ignition rotor points to the number 1 cylinder position, both valves should be closed. Valves that are open, just closing, or just opening indicate incorrect valve timing. It should be mentioned here that if the error in valve timing is due to "jumping a tooth," ignition timing will also be incorrect if the distributor is driven by the camshaft.

A more exact test can be made at TDC on the exhaust stroke. Because this is the valve overlap position, both valves should be slightly open. The exhaust valve should be just closing, and the intake valve should be

*FIGURE 2-6 A camshaft sprocket that has been badly worn by the timing chain*

just opening. Both valves should move simultaneously when the crankshaft is "rocked" (turned slightly forward and backward).

Since excessive backlash is likely to occur, cam drive backlash should be inspected whenever valve grinding or overhauling is considered. Because of changes in valve timing, a high backlash condition does not allow an engine to run with normal power and economy. More importantly, the timing chain and sprockets commonly fail, causing erratic engine operation or engine failure (including bent valves). When backlash exceeds 5°, replacement of the cam drive sprockets and camshaft timing chain should be included in any repair estimate for valve grinding. In engines that have high mileage it is not uncommon to find from 10° to 15° of backlash.

Excessive backlash can be confirmed in many cases by testing for timing chain backlash as follows:

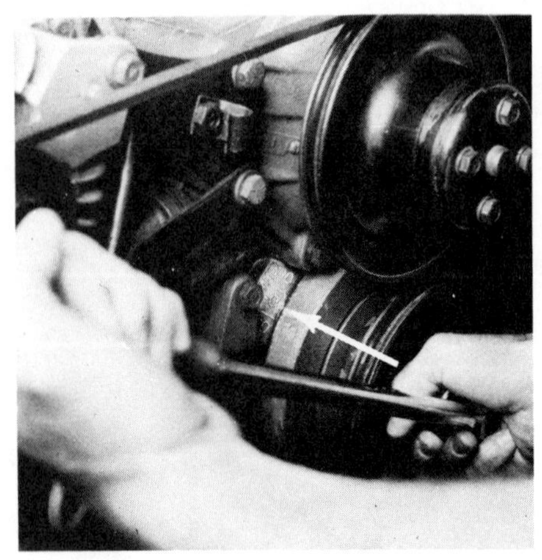

*FIGURE 2-7 Turning the engine forward to TDC*

1. Turn the engine forward until the timing marks are at TDC (see Fig. 2-7).
2. Mark the position of the rotor. Turn the engine backward by hand until the rotor just begins to move (see Fig. 2-8).
3. Read the amount of backlash at the timing marks.

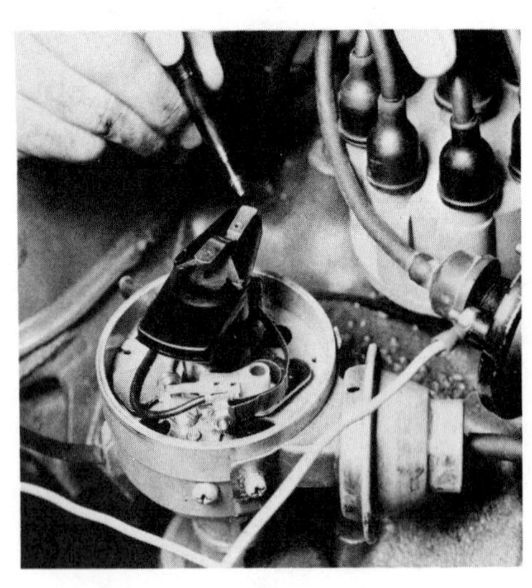

*FIGURE 2-8 Watching for motion at the ignition rotor*

## MANIFOLD VACUUM TESTS

Testing the manifold vacuum may also uncover engine defects. This test is performed as a part of tune-up procedures or as a follow-up to a compression test. First, it is important that the concept of a "normal" vacuum gauge reading be understood. Normal vacuum in the manifold varies with the altitude and with the engine design. Gauge readings must be com-

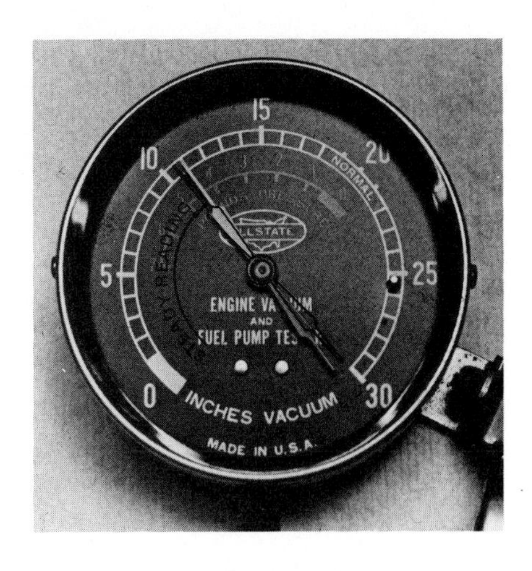


*FIGURE 2-9 A low and steady vacuum gauge reading*

*FIGURE 2-10 Intermittent drop in a vacuum gauge reading*

pared to what is expected from a sound engine of a certain design, at a specific altitude (the vacuum readings will drop approximately 1 inch for every 1,000 feet above sea level). The vacuum gauge should be connected with as short a hose as possible so that the needle can respond quickly to changes in pressure. Some gauge readings and their associated valve train defects are:

1. Low and steady readings caused by manifold leaks, late ignition timing or possibly by worn piston rings (see Fig. 2-9).
2. Intermittently dropping readings caused by sticking valves (see Fig. 2-10).
3. Rapid needle oscillation at idle caused by leaking valves (see Fig. 2-11).
4. Rapid needle oscillation on acceleration (similar to that illustrated in Figure 2-11) caused by weak valve springs.

The test procedure used to diagnose piston ring leakage is as follows:

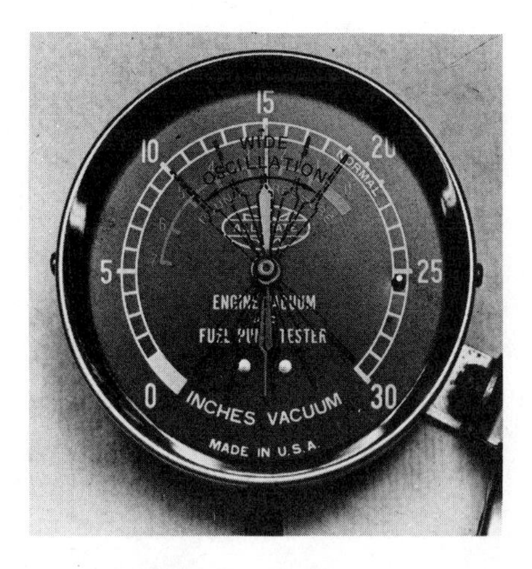


*FIGURE 2-11 Rapid needle oscillation on a vacuum gauge*

*FIGURE 2-12 A normal vacuum gauge reading of 17 to 21 inches of mercury, depending on the engine design*

*FIGURE 2-13 A normal increase in vacuum when the throttle is closed*

1. Connect the gauge and observe the "normal" reading at idle (see Fig. 2-12).
2. Increase the engine speed to approximately 2,000 RPM.
3. Close the throttle rapidly and observe the gauge reading. An increase of less than 2 inches of mercury (in. Hg) would indicate piston ring leakage (see Fig. 2-13).

Again, it is emphasized that the mechanic must know what "normal" is. Note that the mechanic must observe the gauge reading and the type of needle action or response. The need for a tune-up, especially ignition timing and valve adjustment, and manifold vacuum leaks must be eliminated as the causes of certain vacuum gauge readings.

*FIGURE 2-14 A stethoscope used to locate the source of an engine noise*

## ENGINE NOISES

On occasion, engine noise alone indicates that engine repair is required. However, some engine noises may be characteristic of a particular engine and not necessarily a probable indication of engine failure. The experience of a skilled mechanic is required to evaluate the seriousness of engine noises (see Fig. 2-14).

Some of the more common engine noises and conditions are listed below:

1. Hydraulic valve lifters are a common source of noise complaints. The noise is heard as a knock occurring at camshaft speed (one-half of engine rpm). The faulty lifter may be isolated by pressing a hammer handle downward on the push rod end of each rocker arm while the engine is idling. The noise will go away or at least be diminished when pressure is applied to the faulty lifter. Before condemning the "faulty" lifter, be sure to check for valve spring, push rod, or rocker arm damage and even valve adjustment because any one of these problems may cause noises similar to defective lifters.
2. Main bearing noises caused by excessive clearance are heard as deep metallic knocks. The noise will be most audible when the engine is under load, accelerating, or just started. Low oil pressure may accompany excess main bearing clearance.
3. Crankshaft end play beyond acceptable limits can be heard as a sharp metallic rap. The noise will be most audible when releasing or engaging the clutch on manual transmission cars. This condition may be tested for by placing a dial indicator at one end of the crankshaft and prying against the crankshaft to force it to its limits of travel (see Fig. 2-15).
4. Connecting rod noises caused by excessive clearance is heard as a light metallic rap when the engine is running under a light load. The noise becomes louder and increases in frequency as engine speed is increased. The defective connecting rods or bearings may be isolated by grounding one spark plug at a time. The engine noise will be noticeably decreased when the spark plug for a cylinder with excess rod bearing clearance is grounded.

*FIGURE 2-15 A dial indicator used to check crankshaft end play*

5. Piston "slap" is a noise typically associated with excess piston-to-cylinder wall clearance. Piston slap is heard as a dull metallic rattle at idle and under light engine loads. Many of these noises disappear when the piston expands during warm-up and in such cases are not a concern as far as engine reliability is concerned. When noise remains after warm-up, piston or ring failures are possible. The particular cylinder, or cylinders, may be isolated by grounding the spark plugs as with locating rod bearing noises.

6. The noise caused by excessive piston pin clearance is heard as a light metallic rap at idle and at low speeds. The noisy piston pin can also be isolated by grounding spark plugs, although the effect of grounding the spark plug is different. That is, when the spark on a cylinder with excessive piston pin clearance is grounded, the noise changes—usually increasing in frequency. This noise is sometimes heard immediately after piston rings are replaced in an engine overhaul. The noise will usually diminish and disappear as friction between the piston rings and cylinders decreases during break-in.

Keep in mind that some noises may be coming from sources other than internal engine parts. Drive belts, alternators, compressors, air pumps, and fuel pumps all make noise. It is sometimes a good idea to run the engine after disconnecting all drive belts and removing the fuel pump from the engine block. The noise in question just might disappear. To prevent overheating, just be sure to run the engine for only a few minutes. Check for damaged flywheel covers also because they may rub against rotating parts. Even flywheels will occasionally be found to be knocking as a result of being loose.

# 3

# Engine Disassembly

Procedures for the disassembly of engines vary from model to model. For this reason, service references should be checked prior to beginning disassembly. However, there are points that are generally common to engine service procedures, and these points are reviewed in this part.

It is also true that the economics of repair may determine to some extent the repairs to be made as well as the procedures to be followed. For example, perhaps it is decided that engine block repairs are to be made after the cylinder heads have been removed for valve service. In such a situation, replacing piston rings and crankshaft bearings with the engine block mounted in the chassis would be a serious consideration. So long as engine service requirements are not too extensive, such a repair procedure could save several hours of labor costs. On the other hand, if the repair called for is a major engine overhaul or engine rebuilding, removing the engine from the chassis and following complete service procedures are required.

Parts should be handled carefully and stored so that they will not be lost. Consider that head bolts, for example, are special high strength fasteners that will not necessarily be easily replaced if lost. Careful handling pays;

if serviceable parts are damaged, additional machining or parts replacement will be necessary. It is not unusual to have to resurface cylinder heads because of gouges or scratches picked up in handling, even though surfaces may remain within limits for flatness.

In general, it is best to follow good mechanical practice. Work according to correct procedures, keep engine parts clean and organized, and handle parts carefully. These basic work habits are a part of the craftsmanship essential to quality engine work.

## HINTS FOR DISASSEMBLY IN THE CHASSIS

As mentioned, engine repairs are frequently performed with the engine mounted in the chassis. This procedure is used to make repairs more economical by reducing labor costs. Of course, valve service is typically done with the block in the car. Examples of other repairs include the replacement of a single defective piston or the replacement of piston rings in one or more cylinders. Simple ring and valve jobs are also often done in this manner. The valves are ground, piston rings are replaced, and crankshaft bearings are replaced.

Disassembly for valve service or other repairs is generally begun by removing the cylinder heads from the engines. This, of course, also involves removing the intake manifold and either removing exhaust manifolds or disconnecting exhaust pipes from the manifolds. An air conditioning compressor can sometimes be tied against a fender well with wire so that it will not have to be disconnected from the system.

If it is also necessary to remove the oil pan from the engine, it is frequently required that steering linkages be disconnected and dropped so that the oil pan may be lowered from the engine block. It is sometimes required that the engine mounts be unbolted and the engine block raised a few inches. It may even be necessary to loosen the pan and unbolt the oil pump before the oil pan can be removed completely. Once the cylinder heads and oil pan are removed, it is possible to begin removing piston and connecting rod assemblies.

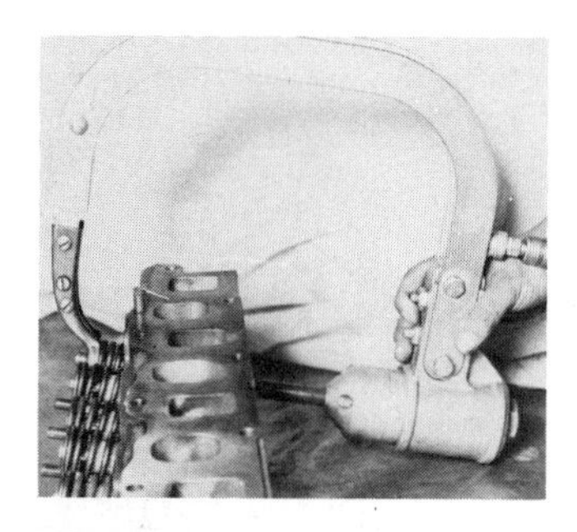

*FIGURE 3-1 Using an air-operated valve spring compressor*

## DISASSEMBLING CYLINDER HEADS

Cylinder head disassembly is done with the aid of spring compressors (see Fig. 3-1). The steps are routine except for a few special cautions.

First, it will be found that valve spring retainers will sometimes "lock up," or seize, on the valve keepers. This is caused by the wedging action of their matching tapers. When locked, valve springs cannot be compressed to remove the keepers. They can be loosened by placing a socket or an old piston pin against the retainer and rapping it with a soft-faced hammer. Be careful to strike straight down along the centerline of the valve stem, or the valve will bend (see Fig. 3-2). Also be sure that the valves do not hit anything when they pop open because this could bend a valve, too.

*FIGURE 3-2 Striking the spring retainer to free it from the valve keepers. Note the use of an old piston pin for a driver.*

Second, it will be found that the tips of valve stems will occasionally "mushroom" and prevent the valve from slipping through the valve guide. The tip of the valve stem may be cleaned with a file or an air grinder (see Fig. 3-3). Don't rush the job by attempting to drive the valve through the valve guide. The result will be a ruined valve guide and additional time loss in repairing it.

*FIGURE 3-3 Filing the mushroomed tip of a valve stem*

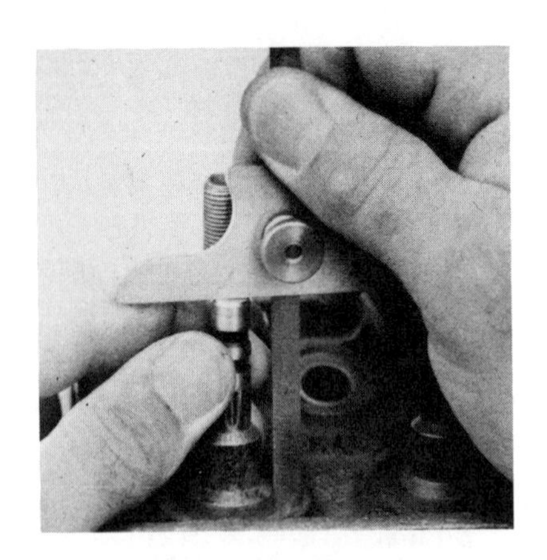

*FIGURE 3-4 Measuring from the tip of the valve stem to the spring seat*

Third, measure the valve stem height while valves are still in order of assembly (see Fig. 3-4). This is critical on engines with non-adjustable rocker arms, because valve stem length changes the position of the plunger in hydraulic valve lifters. After reassembly, the length must be the same as before disassembly, or valves may be held open. Specifications for valve height are not always available, because the manufacturers frequently use special fixed gauges (see Fig. 3-5). Unfortunately, these fixed gauges are not on hand in most independent shops.

Over the years, it has become standard procedure to keep valves in order for reassembly. Unless oversize stems are used, there is minimal value in this practice since all valves and valve guides must be measured for wear and checked against specifications. Keeping them in order during disassembly might be useful only in that if parts with defects are noticed, the mating valve guides or rocker arms can be checked for damage. Clearances when reassembled will be changed very little because only parts measuring within service limits will be reused.

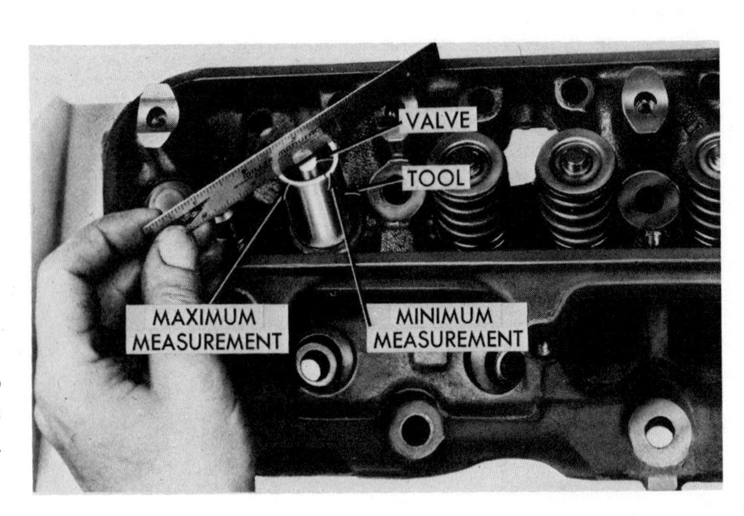


*FIGURE 3-5 A fixed gauge used to check valve stem height above the head (Courtesy of Chrysler Corp.)*

## NUMBERING CONNECTING RODS

There are checks and procedures to follow prior to unbolting connecting rods. For example, check to see that connecting rods are numbered according to cylinders; they may or may not be numbered at the factory. A number corresponding to the cylinder should be stamped on each half of the connecting rod (see Fig. 3-6). The numbers are usually stamped all on the same side of in-line engines or on the same side as the cylinder on V-block engines (see Fig. 3-7). Check carefully because the rod numbering, if present, could be incorrect. If the rods are not numbered, they should be stamped while they are still on the crankshaft.

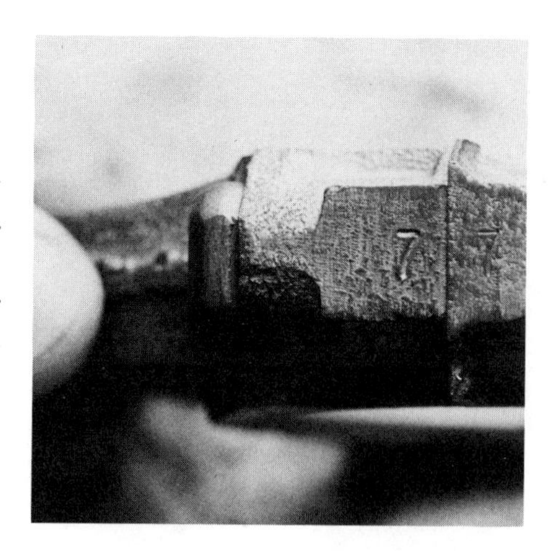

*FIGURE 3-6 Cylinder number stamped on both sides of the parting line of each connecting rod*

Numbering them in this manner does several things. First, stamping both—the connecting rod and the cap—aids the mechanic in replacing the cap on the rod in the original position. If the cap should be replaced on the wrong rod or in the wrong direction, the housing bore of the connecting rod will be out of round beyond acceptable limits.

*FIGURE 3-7 Numbering connecting rods*

Second, stamping all connecting rods on the same side of in-line engines, or on the same side as the cylinder on V-block engines aids the mechanic in replacing piston and connecting rod assemblies in the cylinder with the piston facing in the proper direction. Valve interference may occur if the piston should be replaced in the cylinder in the wrong direction (see Fig. 3-8).

Third, stamping the connecting rods while they are assembled to the crankshaft minimizes any distortion that may occur as a result of the stamping. Instead of stamping, it is possible to number the connecting rods with an

*FIGURE 3-8 Points to observe to ensure correct installation of connecting rods and pistons (Courtesy of Chrysler Corp.)*

electric pencil or similar tool; however, it is still recommended that connecting rods be numbered before disassembly to eliminate confusion as to their direction or position.

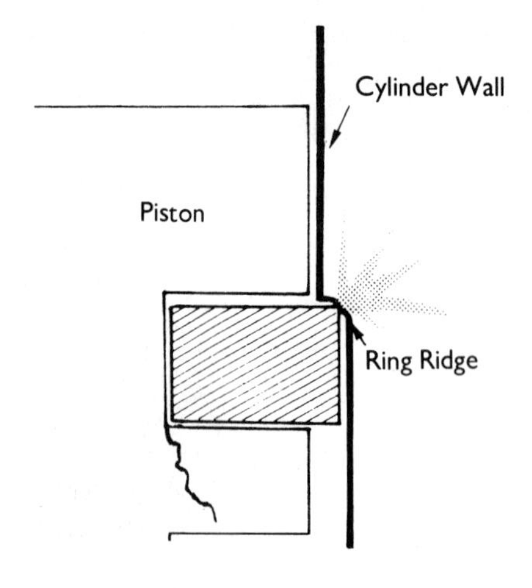


*FIGURE 3-9 Piston ring striking the ring ridge*

## RIDGE REAMING

Cylinder wear is generally concentrated in the top inch of piston ring travel; therefore, a ridge forms at the top of the piston ring travel. That ridge may well cause difficulty during engine disassembly. Unless reboring is to be done, this ridge must be removed to prevent possible damage to piston ring lands when piston and connecting rod assemblies are removed. Damage to ring lands occurs as pistons are pushed out of the cylinder and piston rings catch on the ring ridge (see Fig. 3-9).

Failure to remove a ring ridge may result in the unnecessary replacement of a piston

Tools called ridge reamers are available for removing the ring ridge (see Fig. 3-10). Care must be taken in this operation to see that ridge reaming does not extend into the area of ring travel. Ridge reaming that extends into the area of ring travel destroys the surface area required for sealing the replacement piston rings.

Because of the irregular wear patterns in the top of the cylinder, it is common to find the ring ridge cannot be fully removed without reaming into the piston ring travel. However, partial ridge reaming will at least reduce the ring ridge enough to make piston and connecting rod removal easy and minimize damage. In every case, minimum force should be used in removing the assemblies.

*FIGURE 3-10 Using a ridge reamer*

*FIGURE 3-11 Crankpin damaged by rod bolts during disassembly*

## REMOVING PISTON AND ROD ASSEMBLIES

It is all too common to find that otherwise serviceable crankshafts are damaged during the removal of piston and connecting rod assemblies (see Fig. 3-11). The cause of the damage is failure to keep connecting rod bolts from contacting crankshaft throws when the connecting rod bolts slide past the crankshaft. This damage is easily prevented by placing a length of rubber tubing over each connecting rod bolt during disassembly (see Fig. 3-12).

Remove one rod cap at a time, and place tubing over the rod bolts. The rod and piston assembly is pushed or driven out through the top of the block. A hardwood dowel or a hammer handle may be used as a driver so that the rod will not be dented, gouged, or otherwise damaged. *Do not attempt to remove assemblies through the bottom side:* they will

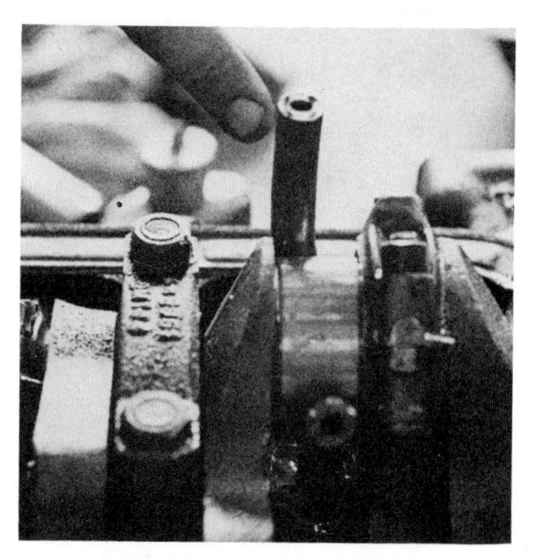

*FIGURE 3-12 Using rubber tubing on rod bolts to protect the crankshaft*

wedge against the main bearing webs, and attempts to move them in either direction will probably lead to broken pistons and damaged rods.

## REMOVING THE TIMING CHAIN AND SPROCKETS

Timing chain slack or timing gear backlash should be checked prior to removing the camshaft or crankshaft from the engine. When timing chain slack is under one-half inch, the timing chain and sprockets are considered by many mechanics to be serviceable (see Fig. 3-13). The wear limit generally accepted for backlash between timing gears is .006 inches (.15 mm) (see Fig. 3-14). It is also true that many mechanics will automatically change sprockets and chains during overhaul. If the slack is excessive, both sprockets and chain are replaced. Use care not to twist a serviceable timing chain when removing it.

*FIGURE 3-13 Checking timing chain slack*

*FIGURE 3-14 Checking timing gear backlash with a feeler gauge*

## REMOVING THE CRANKSHAFT

*FIGURE 3-15 One method of marking main caps*

As with connecting rods, the main bearing caps are not always marked for position or direction. The caps may be marked with numbers on one side and the engine block marked at corresponding points (see Fig. 3-15). Main bearing caps that are repositioned in the wrong location or in the wrong direction will cause main bearing housing bores to be misaligned and out of round beyond acceptable limits.

Once the main bearing caps are marked, they may be removed and set aside. The timing chain should be removed by this time also. The crankshaft can now be lifted out of the crankcase and stored on end. Crankshafts are stored on end to prevent the distortion that can occur when they are stored lying flat. The main bearings should also be removed at this time.

Another suggestion is to replace the main bearing caps on the block after removing the crankshaft and bearings. This prevents misplacing the bearing caps or bolts (capscrews). It is also suggested that block machining operations be performed with the main bearing caps tightened in place so that engine block distortion caused by tightening capscrews will be the same as in the assembled engine. Observing this practice will ensure that machining accuracy will be best in the assembled engine rather than in the disassembled one.

## REMOVING THE CAMSHAFT AND VALVE LIFTERS

In an engine overhaul, as many parts as are serviceable are reused to make the repair more economical. Therefore, it is important that the valve lifters be kept in order so that they can be replaced in their original locations. The valve lifters should be stored in a box or possibly in a wooden block with holes drilled in it so that their positions can be marked. *If valve lifters are replaced in incorrect positions, camshaft failure will be almost immediate.*

Valve lifters are sometimes difficult to remove because of varnish buildup on the portion of the valve lifter extending below the lifter bores. Removal can be made easier by spraying penetrating oil around the valve lifters and lifter bores as soon as they are exposed. Small tools are also available for pulling the valve lifters out. The valve lifters can also be pushed through the lifter bores *from the top* with a push rod after the camshaft is removed. *Never push lifters from the bottom* because damage to the lifter bores will result.

Check to see whether a camshaft thrust plate is used on the engine to limit camshaft end play (see Fig. 3-16). Thrust plates, if used, must be loosened before attempting to remove the camshaft. The retaining screws can usually be readily removed through holes in the timing gear or sprocket.

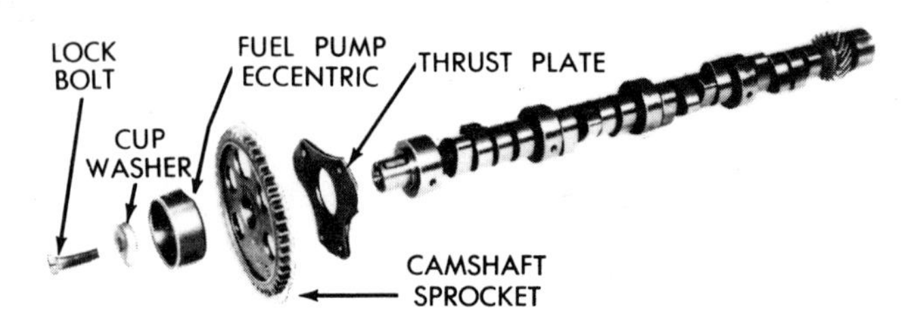


*FIGURE 3-16 A camshaft thrust plate in its relative position with camshaft components (Courtesy of Chrysler Corp.)*

## REMOVING CAMSHAFT BEARINGS

Various manufacturers make tool sets for removing and installing cam bearings. It is important to select drivers properly or to adjust them (see Fig. 3-17) so that bearing bores in the engine block are not scored during the process of removing bearings (see Fig. 3-18). Surface damage to these bores, if not corrected, will cause the new cam bearings to distort or to misalign. Extra care taken at this point will also make engine assembly easier. In fact, efforts to remove cam bearings with improper tools (chisels, punches, etc.) can cause the block to be scrapped.

*FIGURE 3-17 Parts of an adjustable driver from a driver set*

## REMOVING OIL PLUGS AND CORE PLUGS

Oil passages in engine blocks are plugged at least at one end with pipe plugs (see Fig. 3-19). This is usually where oil will leak externally if

*FIGURE 3-18 Driving out a cam bearing*

*FIGURE 3-19 Pipe plugs used to close one end of oil passages*

not well sealed, although they may be used internally as well. It is necessary to remove these plugs so that oil passages may be scrubbed clean. Keep in mind that hot tank cleaning of the engine block will loosen deposits in passages and severe scoring of the new engine bearings will result if passages are not thoroughly cleaned. In fact, solid contaminants left in oil passages are the number one cause of bearing damage.

The pipe plugs are most easily removed by heating them red hot with an oxyacetylene torch. When cooled, the threads of the pipe plugs are sprayed with penetrating oil, and the plugs are removed easily with a wrench. It is wise to steam clean the engine block first to minimize fire hazards.

One end of each oil passage is usually also plugged with a small diameter expansion or core plug (see Fig. 3-20). This is usually where oil will leak internally if not well sealed. These coreplugs are easily driven out with a ¼ inch (6 mm) diameter rod inserted from the opposite end of the passage once the pipe plugs are removed. Another good practice is

*FIGURE 3-20 Core plugs in oil passages*

to measure the core plug on removal and note the size so that replacements can be ordered.

Core plugs for water jackets are also removed for hot tank cleaning. They are removed even if relatively new because the hot tank solution removes surface plating and the rate of corrosion is accelerated. One quick way of removing them is to use a punch to knock them through to the interior of the block (see Fig. 3-21) and a pry bar to pull them back out (see Fig. 3-22).

The core plugs must not be left in the water jackets, or they will disrupt the flow of coolant. As with cam bearing bores, care must be taken not to damage the bore, or the replacement plugs will not seal.

*FIGURE 3-21 Driving a core plug inside the water jacket before removing it*

*FIGURE 3-22 Using a rolling head pry bar to remove a core plug*

# 4

# Cleaning Engine Parts

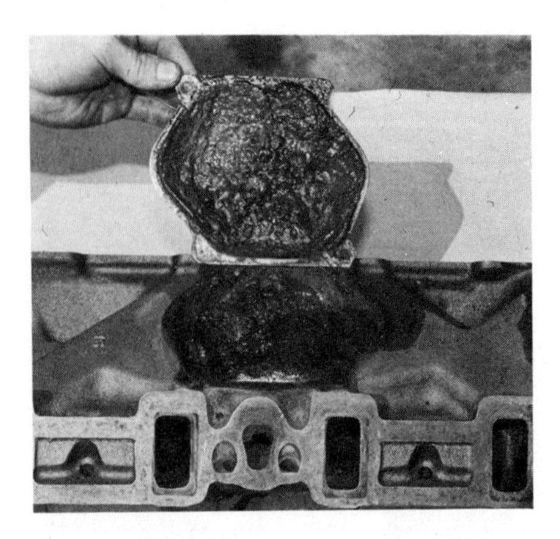

*FIGURE 4-1 Carbon deposits hidden under an intake manifold*

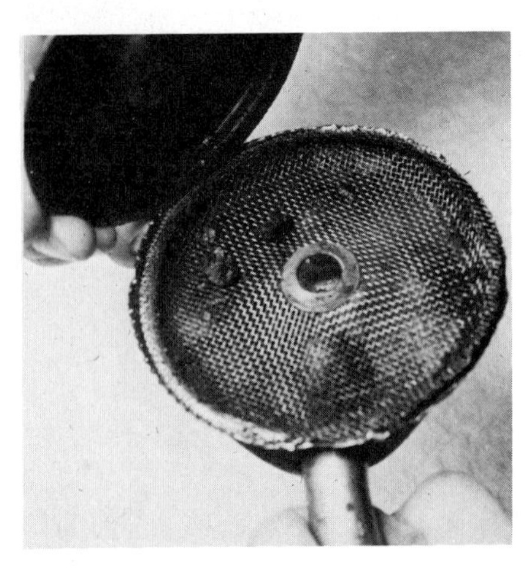

*FIGURE 4-2 Opening a pick-up screen to remove dirt*

Parts cleaning is an essential part of quality engine service. Cleaning, however, is also a time-consuming process, and there is a temptation to shortcut cleaning steps. A careful selection of procedures should be made so that cleaning may be completed as thoroughly and as quickly as possible.

A major cause of poor performance in reconditioned engines is solid contaminants in valve guides and engine bearings. Deposits collect in engines for 100,000 miles or more without causing problems, but they are loosened by cleaning solutions and are carried throughout the engine after servicing.

One common source of hidden dirt will be found under the floor of V-block intake manifolds (see Fig. 4-1). To clean this area, remove the pins holding the shield in place by wedging a cold chisel under the head of each pin. The shield can be replaced by reinstalling pins or by drilling and tapping for screws.

Another commonly overlooked source of dirt is the oil pump pickup screen. It may be soaked, scrubbed, or flushed, but very probably it will remain dirty unless taken apart for cleaning (see Fig. 4-2). If the pickup cannot be pryed open for cleaning, it is recommended that it be replaced.

## CLEANING SOLVENT

*FIGURE 4-3 Using solvent to clean an oil passage. The solvent is pumped through the nozzle*

The most common cleaning solution available to mechanics is cleaning solvent. Solvent is used cold for small degreasing jobs, and generally some hand scrubbing and scraping is also necessary. Solvent tanks are available with pump and agitation systems for improved efficiency (see Fig. 4-3). Cleaning solvent may be used for all metallic materials used in automotive engines.

## COLD TANK SOLUTIONS

Solutions are available for the cold-soaking of engine parts. These may be general purpose cleaners for use with ferrous metals (usually iron and steel) and nonferrous metals (usually aluminum), or they may be used for more specific cleaning tasks. Parts are left to soak in these solutions until clean and then rinsed in water.

## HOT TANK SOLUTIONS

Hot solutions clean considerably faster than cold solutions. Temperatures of 160°F or more are used. As with cold solutions, the available chemicals (usually alkaline) may be used with both ferrous and nonferrous materials, or they may be used with only one of the two. Because of the severity of these solutions, it would be a serious mistake to place parts in the wrong solution. For example, cam bearings will be ruined by ferrous solutions. If in doubt about the type of metal, check for ferrous metals with a magnet.

As mentioned, these solutions may be used with ferrous *and* nonferrous metals or with ferrous *or* nonferrous metals. It is recommended that engine materials be considered carefully when selecting solutions. For example, aluminum-magnesium alloys are sometimes cleaned with "aviation" solutions rather than solutions more common to automotive use. The manufacturers of the solutions are generally cooperative in making recommendations for special job requirements.

Be sure not to place assemblies, such as pistons and rods, in these solutions because chemical deposits between moving parts will cause assemblies to bind. It is also recommended that particular care be used with valve springs being degreased in hot tank solution. Parts removed from these solutions are sometimes so thoroughly degreased that corrosion readily takes place after they are removed from the hot tank. Springs should be rinsed and sprayed with corrosion inhibitor as quickly as possible. Valve springs that have become corroded should be discarded because of the possibility of stress concentration and subsequent failure.

*FIGURE 4-4 A spray jet hot tank for degreasing engine parts.*

A variety of hot tank equipment is available. The variations include tank size and the presence of agitation devices. For example, many small nonproduction shops use hot tanks without parts agitation or spray jets. Production shops, however, seek the speed and efficiency gained by the addition of parts agitation or spray jet systems (see Fig. 4-4).

## DESCALING

Occasionally it is found that the water jackets in cast-iron cylinder heads and engine blocks are plugged with rust and scale. This is especially true of marine engines. Such deposits inhibit the transfer of heat from the engine parts to the coolant. Unless these rusty or scaly deposits are removed, engine damage from overheating results.

Cold acid descalers can be used to remove rust from iron castings. The parts are degreased and rinsed thoroughly before soaking them in the descaler. The parts are removed from the descaler solution after approximately an hour and again rinsed thoroughly. Care must be used when working with these solutions to prevent splashing the solution onto skin or clothing.

It is also important that rinsing after degreasing be as thorough as possible to prevent any carryover of the alkaline degreaser into the acid descaler. Such contamination would cause a chemical neutralization of the acid descaler. It would tend to be changed to water and salt. Because of the expense of these solutions, careful adherence to procedures is a must.

## BEAD BLASTING

Carbon removal on many engine parts is a particular problem. The various solutions used for cleaning do not fully remove carbon because they are intended primarily for degreasing, in which case carbon removal is limited to deposits held together by oily materials. The carbon collects on those engine parts directly exposed to hot combustion gases.

Bead blasting with glass shot (see Fig. 4-5) is a process now commonly used for decarbonizing cylinder head combustion chambers and ports and valves. It has further application in decarbonizing other engine parts and in removing rust or gasket material. The process saves considerable time otherwise spent hand scraping and wire buffing.

*It is recommended that bead blasting not be used where abrasive can enter threaded holes or oil passages* that are difficult to clean. It is also recommended that it not be used on piston skirts that have antiscuff coatings (tin or iron oxide).

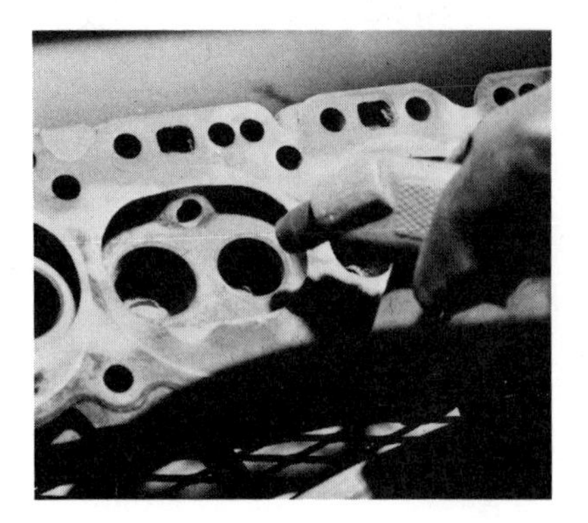

*FIGURE 4-5 Beadblasting engine parts*

*FIGURE 4-6 Tumbling a cylinder head inside truck tires to remove beads*

Again, it must be emphasized that beads must be kept out of, or removed from, threaded holes, oil passages, and the interior of the engine. Beads should be removed from holes with compressed air and bore brushes. Water jackets may be flushed with water or a steam cleaner. One production method of removing beads is shown in Figure 4-6. Failure to remove beads causes problems with threads, engine wear, and water pump failures.

## HAND AND POWER TOOLS

Much cleaning is done with small hand and power tools. A gasket scraper is a wide, sharp-edged tool suitable for removing tightly bonded gasket sealer and material (see Fig. 4-7). Ring groove scrapers may be adjusted to fit the various standard ring groove widths and depths (see Fig. 4-8) so that carbon can be removed without damage to the base metal.

Oil passages are often cleaned with bore brushes. Bore brushes are available in different diameters and lengths suitable for engine block

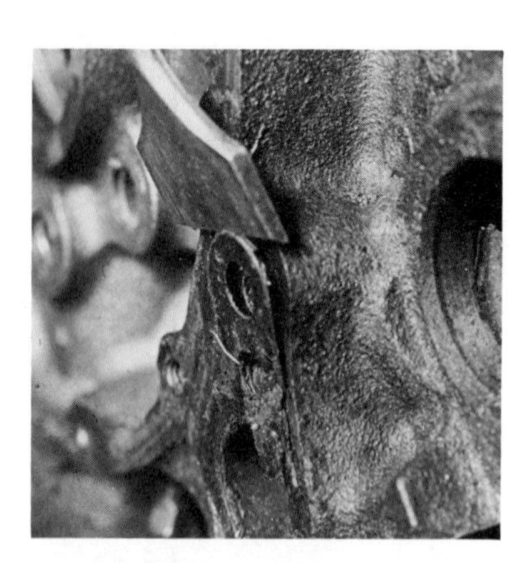

*FIGURE 4-7 Using a scraper to remove an old gasket*

*FIGURE 4-8 Using a ring groove scraper*

oil passages, valve lifter bores, or crankshaft oil passages.

A wire buffer is a convenient tool for the last minute removal of gasket material or sealer. The buffer is used as an alternative to hand scraping or bead blasting (see Fig. 4-9).

An air motor and flare brush combination was the forerunner of bead blasting for cleaning combustion chambers and ports (see Fig. 4-10). Although not as fast or thorough as bead blasting, it is the best alternative if bead blasting equipment is not available.

*FIGURE 4-9 Using a wire buffer to clean threads*

*FIGURE 4-10 A flare brush used to remove deposits from a combustion chamber*

# 5

# Inspecting Valve Train Components

Valve grinding has become such a routine procedure over the years that there is often a tendency not to consider the changes in engines. Newer engines are such tremendous heat producers that valve performance simply cannot be ensured without thoroughly detailed inspection and machining during valve service. The valve job that may have been sufficient twenty years ago will not be adequate under the high heat conditions inside the power plants currently manufactured.

## DETERMINING VALVE GUIDE WEAR

*FIGURE 5-1 Using a telescoping gauge to determine valve guide wear*

Valve guides must be checked for wear, called "bellmouthing," which occurs at each end of the valve guide. It is generally most severe on the combustion chamber side of the valve guide because of the higher temperature and abrasive deposits carried into the valve guide on the valve stem. Valve guide wear can be determined (see Fig. 5-1) by taking measurements of the valve guide diameter in the middle (least worn) and at the combustion chamber end (most worn). The difference in the two measurements will indicate the wear. The measurement of valve guide wear is extremely important in determining what procedures are to be used in restoring specified clearance between the valve guide and the valve stem.

It should be noted that many manufacturers do not give specifications for the valve guide diameter. Instead, the valve stem diameter is specified and the range of guide-to-stem clearance is given. The valve stem must first be measured and compared to specifications. Be aware that the valve stem may already be an oversize diameter. Valve guide diameters may then be measured and compared to the valve stem diameters to determine clearance or wear.

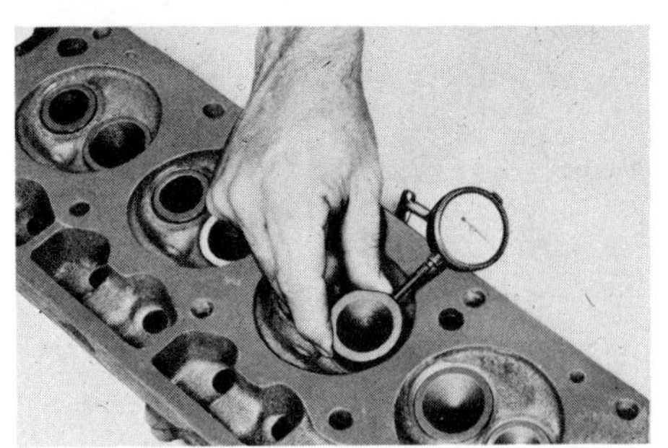

*FIGURE 5-2 Measuring valve rock (Courtesy of Chrysler Corp.)*

Some domestic and imported car manufacturers recommend checking guides by measuring *valve rock.* The valve is opened slightly, and the edge of the valve is "rocked" against a dial indicator (see Fig. 5-2). Because the valve extends out of the guide, valve rock readings will exceed specified guide-to-stem clearance. Be sure not to confuse specifications for guide-to-stem clearance with valve rock when checking valve guides by this method.

Valve guide condition must be carefully checked during routine valve service procedures. It will be found that most cylinder heads will have guide-to-stem clearance exceeding specified limits. Correct sealing at the valve face is virtually impossible with excessive valve guide clearance (see Fig. 5-3).

*FIGURE 5-3 The cocking of a valve at the seat caused by valve guide wear called "bell-mouthing"*

## CHECKING VALVES

It may save time to inspect valves prior to regrinding the faces. Valves that show wear beyond service limits may be discarded before going ahead with machining. Badly damaged valves may be discarded even before cleaning.

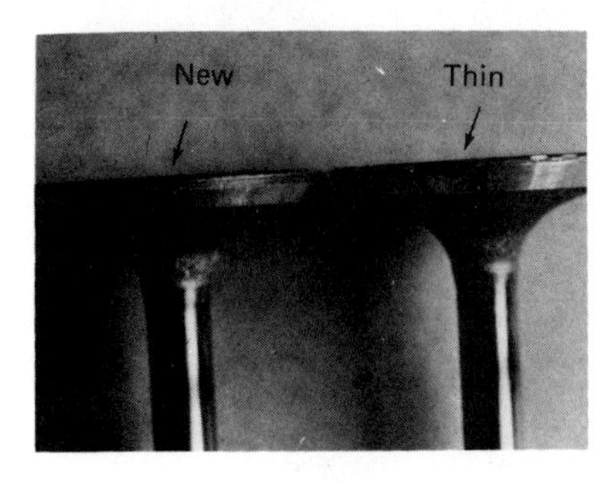


*FIGURE 5-4 The thickness of valve margins*

Check the thickness of valve margins first (see Fig. 5-4). The minimum thickness of the margin after grinding should as a general rule be no less than 1/32 inch (.8 mm) or one-half of new thickness for passenger car engines.

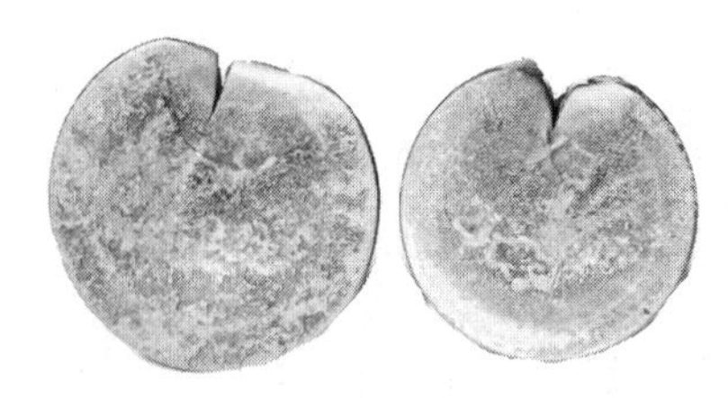

*FIGURE 5-5 Heat-damaged valves*

*FIGURE 5-6 A "necked" valve stem*

*FIGURE 5-7 Wear caused when a rotator fails*

Valves with margins less than 1/32 inch in thickness should be replaced because they would very likely burn up in service. For heavy duty applications, 1/32 inch may well be too thin. Check manufacturers' recommendations if there is any doubt regarding minimum specifications.

Valve stem wear should be measured with a micrometer. A common service limit used is .001 inch (.025 mm) maximum variation in the diameter of the valve stem along its length. Keep in mind that many manufacturers do specify the stem diameter. Valves worn under the specified minimum diameter will be difficult to assemble within specified limits of clearance. Be sure to check specifications because some valve stems are tapered by design to compensate for uneven heating and expansion along the stem.

Valves should also be inspected visually for damage (see Fig. 5-5). Look to see whether the head of the valve is warped, burned, or cupped. Look also for a "necking down" of the valve beneath the head, indicating a stretching of the valve (see Fig. 5-6). These defects are signs of extreme engine heat and material damage that might lead to failure. Some car manufacturers recommend replacing exhaust valves (not regrinding) during valve service partly because of the heat damage which frequently occurs.

Check for grooves worn in the faces of valve stems, usually exhaust because they may indicate defective valve *rotators* (see Fig. 5-7). Valve rotators are a type of valve spring retainer that cause valves to rotate each time they open. Valve rotation extends the life of the valve because temperatures are kept more constant. Valve rotators were common for many years in heavy-duty truck engines and are now common in passenger car engines. Replace these rotators when grooves appear in the faces of the valve stems.

Keeper grooves will sometimes wear and should be visually checked for damage. Worn keeper grooves prevent the keepers from properly locking to the valve stem. Sometimes the grooves wear so evenly that they appear as if they were manufactured that way. If in doubt, compare the worn part to a replacement part (see Fig. 5-8).

The importance of maintaining correct guide clearance and carefully checking valves cannot be overemphasized. Considerable heat is transferred through valves, valve seats, and guides into the cooling system (see Fig. 5-9); therefore, valve seats and guides not kept within service limits may cause valves to fail because of overheating.

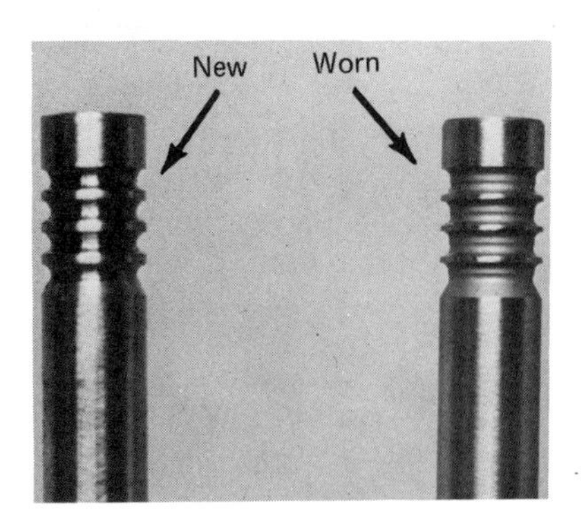


*FIGURE 5-8 Badly worn valve keeper grooves*

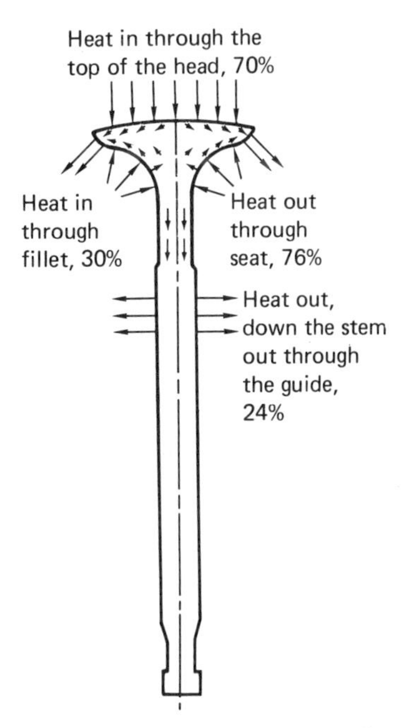


*FIGURE 5-9 Percentage of heat transferred through a valve*

## TESTING VALVE SPRINGS

Valve springs should be tested for tension and warpage. Weak valve springs will fail to close the valve and they may break. A broken valve spring may in turn lead to more severe engine damage by permitting the valve to drop into the cylinder (see Fig. 5-10).

*FIGURE 5-10 Piston damage resulting from a "dropped" valve*

*FIGURE 5-11 Testing spring tension*

Springs should be within 10 percent of specified tension at the given test length (see Fig. 5-11). Test specifications are sometimes given in two sets. That is, one tension is specified for testing with the spring compressed to the "valve closed" length, and another tension is specified for testing with the spring compressed to the "valve open" length. A maximum allowable warpage for valve springs in common use is 1/16 inch (1.5 mm) for each 2 inches (51 mm) of spring length (see Fig. 5-12). Weak or warped springs are showing the first signs of failure.

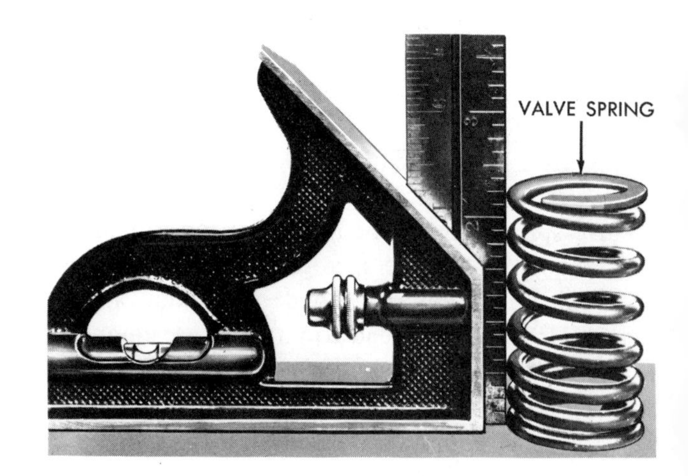


*FIGURE 5-12 One method of checking spring distortion (Courtesy of Chrysler Corp.)*

## INSPECTING THE CAMSHAFT AND VALVE LIFTERS

Hydraulic valve lifters and the camshaft are commonly overlooked in valve service. The valve grinding may be done correctly, but if a valve lifter makes noise or if some other related problem remains, the driver of the car will be dissatisfied. There are inspection procedures that can help to ensure good operation of the entire valve train.

A first check to be made is the visual inspection of valve lifter and cam lobe wear

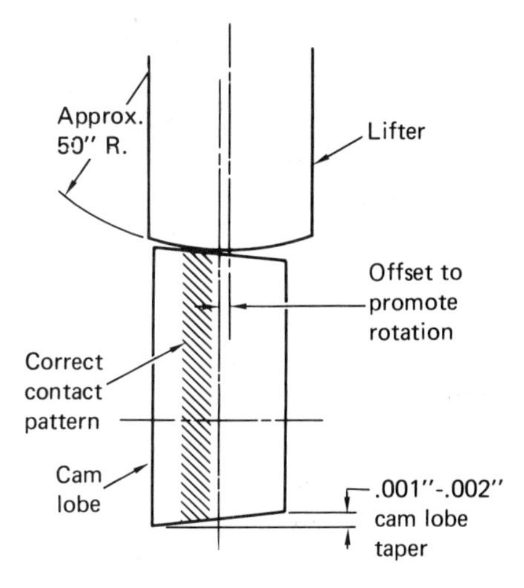


*FIGURE 5-13 The design configuration of valve lifters and camshafts*

*FIGURE 5-14 A badly worn cam lobe*

patterns. To determine how wear patterns develop, consider that new parts are designed to rotate the valve lifter in its bore to minimize wear of the cam lobe and lifter base (see Fig. 5-13).

Occasionally valve lifters will stick in their bores and fail to rotate or rotate only intermittently. This condition is caused by varnish, dirt, or damage to lifter bores and causes extremely rapid wear of the lifter bases and camshaft (see Fig. 5-14). When nonrotation or intermittent rotation is apparent, it is almost certain that both the lifters and camshaft must be replaced (see Fig. 5-15).

Most experienced engine mechanics would be hesitant to make a guarantee of engine life when the valve lifter bases begin to wear in a concave pattern (see Fig. 5-16). Watch also for severe pitting and edge wear patterns on cam lobes, which indicate approaching failure (see Fig. 5-17).

The wear of cam lobes can also be determined by measuring the height of exhaust and intake cam lobes with a micrometer (see

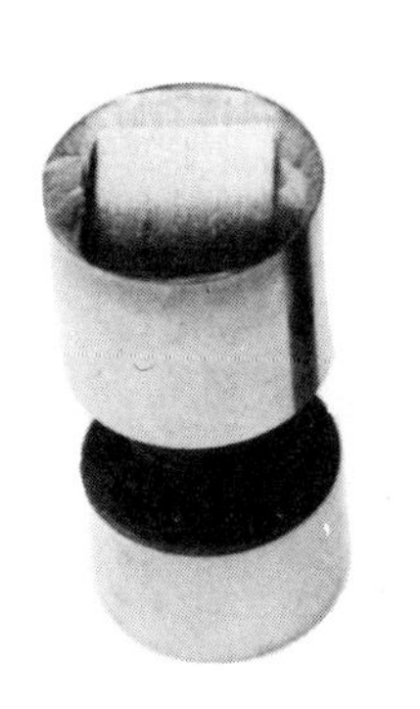

*FIGURE 5-15 Wear caused by nonrotation is shown on the left and wear caused by intermittent rotation is shown on the right.*

*FIGURE 5-16 Badly worn valve lifter bases*

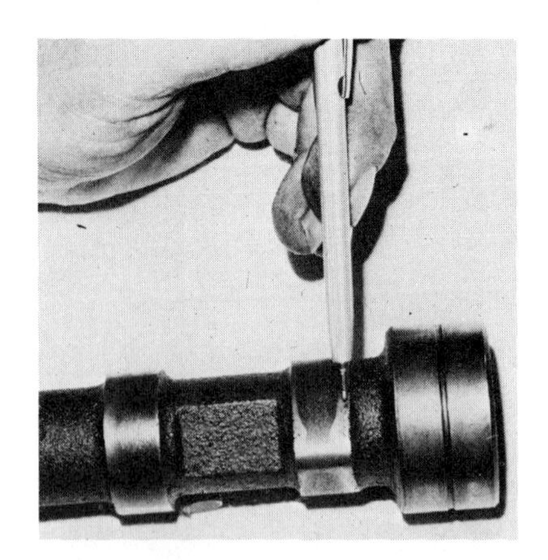

*FIGURE 5-17 Edge wear on a cam lobe*

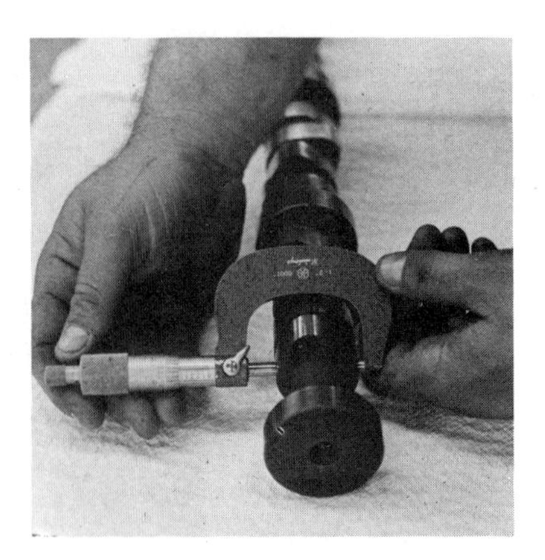

*FIGURE 5-18 Measuring across the base circle of a cam lobe*

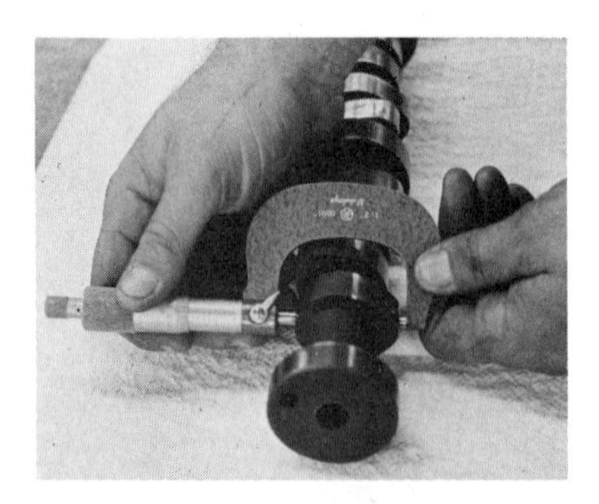

*FIGURE 5-19 Measuring to determine the cam lift*

Figs. 5-18 and 5-19). One recommendation is to consider a variation of .005 inch (.127 mm) between exhaust cam lobes or between intake cam lobes as unacceptable. The diameters of cam shaft journals can also be measured with a micrometer and checked against specifications.

A hydraulic valve lifter leakdown test should also be performed (see Fig. 5-20). If body-to-plunger clearance is excessive or check valves are malfunctioning (see Fig. 5-21), the lifter will collapse almost immediately. This test may also be performed in a small arbor press since the difference in leakdown rates between good and bad valve lifters is obvious —even by feel. Disassembly, cleaning, and retesting lifters is not always recommended because of labor costs and because such procedures do not always cure the problem. Complete sets of valve lifters should be replaced only when the camshaft is also replaced. The reverse is also true; new camshafts should be

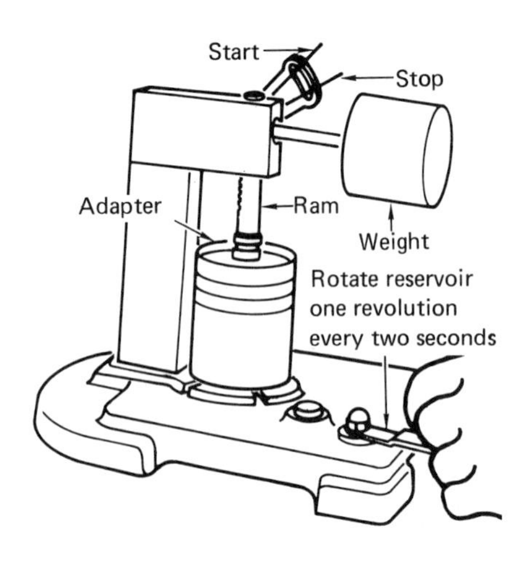


*FIGURE 5-20 Testing hydraulic valve lifter leakdown rates. A defective valve lifter is indicated by a fast leakdown rate (Courtesy of Buick Motor Div.)*

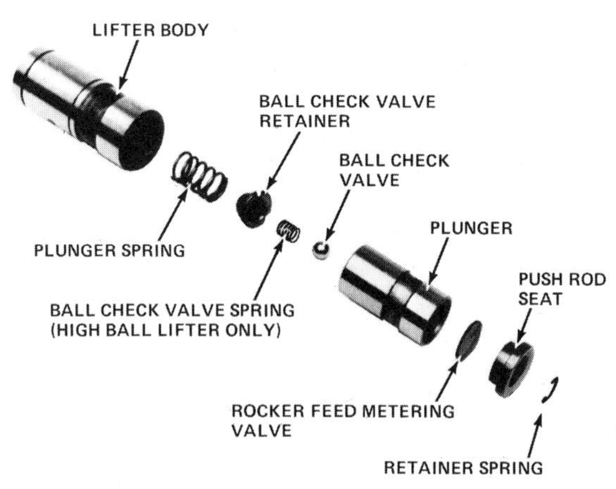


*FIGURE 5-21 The internal parts of a valve lifter (Courtesy of Pontiac Motor Div.)*

installed only with new valve lifters. As a matter of economic necessity, it is recommended that if one defective lifter is found, only the one lifter be replaced.

## CHECKING TIMING CHAIN SLACK OR GEAR BACKLASH

The condition of cam drive components should also be considered. Checking for backlash was covered under "Engine Diagnosis," and in many cases it is the only realistic test because further engine disassembly is not required. However, if valve service happens to be part of an engine overhaul, it is possible to check further. The more complete tests are for timing chain slack (see Fig. 5-22) and for timing gear backlash (see Fig. 5-23). The maximum service limit in general use for chain slack is 1/2 inch (13 mm) although it is recommended here that 1/4 inch (6 mm) be used. For gear tooth backlash, .006 inch (.15 mm) is the limit. If wear is over service limits, *all* cam drive components should be replaced. If replaced, care must be taken to observe the valve timing marks on the gears or sprockets to ensure correct assembly.

*FIGURE 5-22 Timing chain slack*

*FIGURE 5-23 Timing gear backlash*

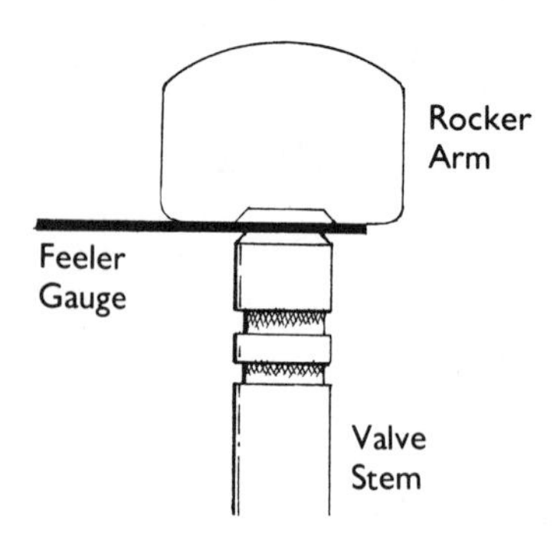


*FIGURE 5-24 Attempting to adjust valve lash with worn rocker arm face*

## INSPECTING ROCKER ARMS AND PUSH RODS

The condition of rocker arm faces is another commonly overlooked point in inspection. Failure to find rocker arm face wear will make the adjustment of valve lash in some engines extremely difficult (see Fig. 5-24). The clearance measured with a feeler gauge will tend to be too loose because the gauge will not slip into the wear pocket (such a situation also tends to ruin the feeler gauge). The poorly matched surfaces of a worn rocker arm placed in assembly with a refaced valve stem will accelerate the rate of wear of both parts.

Particular care should be taken to inspect the so-called *stamped* steel rocker arms. The points to inspect include the push rod socket and the face (see Figs. 5-25 and 5-26). If wear is detectable on the face or in the push rod socket, the rocker arm should be replaced.

*FIGURE 5-25 Badly worn or damaged push rod sockets*

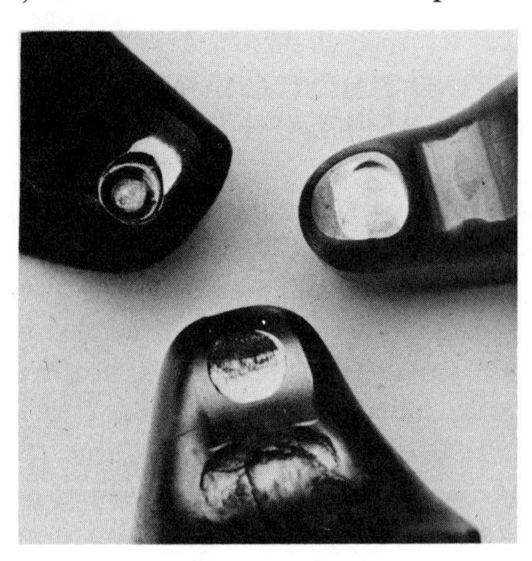

*FIGURE 5-26 Badly worn rocker arm faces*

Badly worn rocker arms will wobble and cut into the sides of the studs (see Fig. 5-27). Also check for variations in height, because the studs will occasionally pull out of the cylinder head (see Fig. 5-28).

*FIGURE 5-27 Badly worn rocker arm stud*

*FIGURE 5-28 Comparing rocker arm stud height with a straightedge*

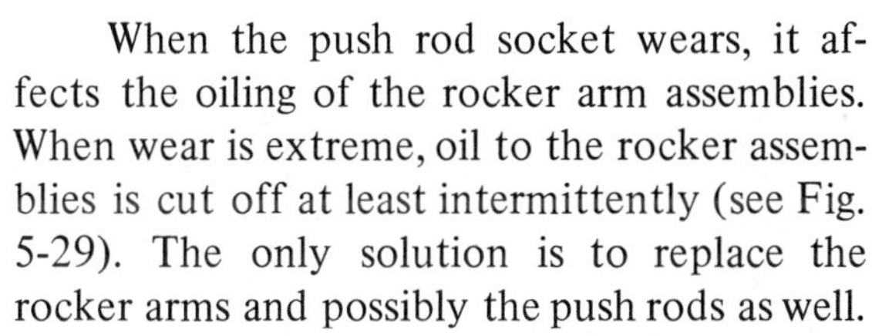

When the push rod socket wears, it affects the oiling of the rocker arm assemblies. When wear is extreme, oil to the rocker assemblies is cut off at least intermittently (see Fig. 5-29). The only solution is to replace the rocker arms and possibly the push rods as well.

The inspections for push rods are relatively simple. The ends must be inspected for any visible wear patterns other than a smooth, round surface (see Fig. 5-30). They should be compared as a set for straightness and for variations in length. Oil passages should be clear. Except for dirty oil passages, which can be cleaned, any adverse conditions discovered call for replacing the worn parts.

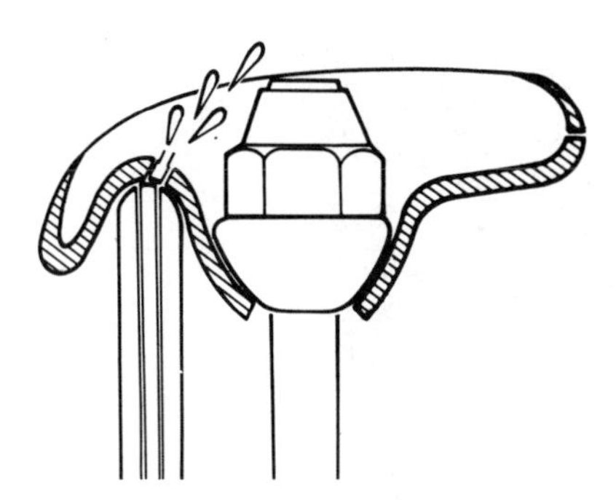

*FIGURE 5-29 Intermittent rocker arm oiling caused by wear of the push rod and push rod socket*

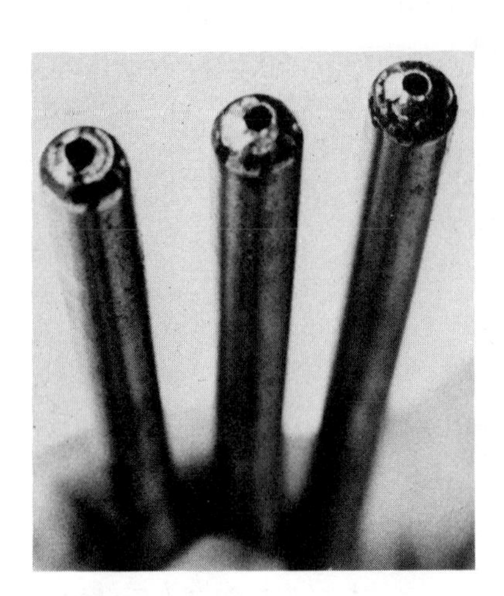

*FIGURE 5-30 The worn tips of push rods*

*FIGURE 5-31 Wear on an overhead camshaft and rocker arm*

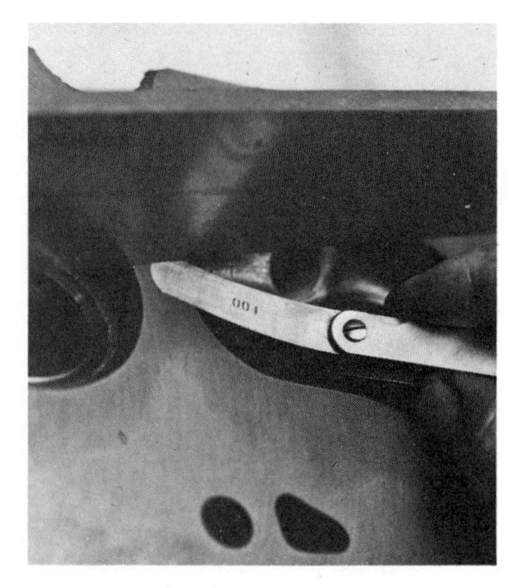

*FIGURE 5-32 Checking cylinder head flatness*

Don't forget that rocker arms for overhead camshaft engines also wear. Give particular attention to wear or scoring on surfaces that contact the camshaft (see Fig. 5-31). Such wear eventually flattens the camshaft and requires that it, as well as worn rocker arms, be replaced.

## CHECKING CYLINDER HEAD SURFACE CONDITION

A good head gasket seal requires a flat, clean gasket surface on the cylinder head (and the engine block, for that matter). The surface of the cylinder head may be damaged from overheating, gasket failure, or perhaps from careless handling.

The check usually thought of first is for flatness. This check is made by using a precision straightedge and a feeler gauge (see Fig. 5-32). A commonly used limit for warpage is .004 inch (.10 mm). That is, when a .004-inch (.10 mm) feeler gauge passes under the straightedge, resurfacing is required. Be sure to check in several directions across the surface (see Fig. 5-33).

Cylinder head surfaces may warp a number of ways. Overhead cam cylinder heads that

*FIGURE 5-33 The directions for checking cylinder head flatness*

warp or bend along their length can cause camshafts to bind in their support bearings. When an overhead cam cylinder head is found to be warped, be sure to check for warpage through camshaft bearings (see Fig. 5-34). Some warped cylinder heads can be salvaged by removing bearing supports and resurfacing the top side of the cylinder head. Others that cannot be *economically* reworked are scrapped.

Check also for damage caused by gasket failure. For this purpose, it is a good idea to keep the original head gaskets on hand and check them for evidence of leakage (see Fig. 5-35). If leakage is apparent at the gasket, check the corresponding point on the head surface. A "blown" head gasket will sometimes permit combustion gases to burn across both cylinder head and block surfaces (see Fig. 5-36).

Corrosion damage around water circulation passages in aluminum cylinder heads will occasionally be severe enough to prevent a head gasket seal. Sometimes corrosion damage

*FIGURE 5-34 Checking for warpage along camshaft bores on the top side of an overhead cam cylinder head*

*FIGURE 5-35 A section of a cylinder head gasket burned away by combustion gases*

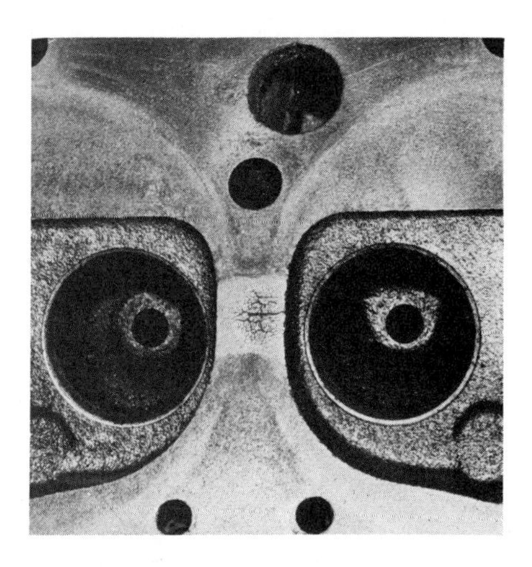

*FIGURE 5-36 The burned or "notched" surface of a cylinder head resulting from head gasket failure*

will extend across oil passages in the head and permit oil to enter the coolant and coolant to enter the oil (see Fig. 5-37). Be sure to check visually for corrosion because such a problem may have been the reason for removing the head. If severe enough, resurfacing will not correct the condition unless damaged areas are built up by welding first.

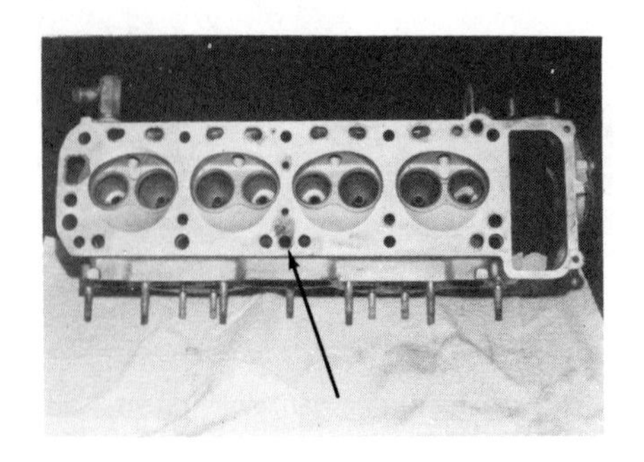

*FIGURE 5-37 Severe corrosion on the surface of an aluminum cylinder head*

As mentioned, surfaces may be damaged merely by careless handling. Look for nicks, dents, or scratches across sealing surfaces. Such defects will cause a gasket failure just as surely as damage caused by severe operating conditions.

Crack detection is also very important, but it involves several other considerations and information on equipment. Because of this, there are separate units on crack detection and crack repair.

# 6

# Inspecting Engine Block Components

Once the engine block is completely disassembled and cleaned, a detailed inspection of all parts should be made to determine their serviceability. It may be found that parts are serviceable as is, that they need reconditioning, or that they need to be replaced.

Engine rebuilding generally implies that the engine is to be restored to new specifications in terms of fits and clearances. This usually requires reboring cylinders and installing new pistons, piston rings, crankshaft bearings, valve lifters, and a new or reground camshaft. It may also require, depending on condition, regrinding the crankshaft, reconditioning connecting rods, and performing other machining operations.

An engine overhaul, however, is approached differently. Measurements of engine parts are made to determine their serviceability. Parts may be worn beyond new specifications but may be considered to be within "service limits." Overhauling is more economical than rebuilding, and an overhauled engine will operate satisfactorily but is not expected to last as long as a new or rebuilt engine.

It is recommended that each engine be disassembled with care and inspected thoroughly to determine wear. An engine may be overhauled for less total cost in labor, machining, and parts. Engine rebuilding should be considered when engine wear is beyond serviceable limits or when it is important to extend engine life to the maximum.

## MEASURING CYLINDER WEAR

Cylinder wear is perhaps the most important consideration in determining whether an engine is within service limits for overhaul. Cylinders are especially important because the expense of reboring and installing new oversize pistons is a major portion of the rebuilding cost. However, if piston rings are replaced and cylinder wear is excessive, ring sealing for any length of time cannot be ensured.

As mentioned in regard to ridge reaming, cylinder wear occurs primarily in the top inch of piston ring travel. Maximum wear is generally found, although not always, across the top of the cylinder perpendicular (90°) to the direction of the piston pin. This is because of the oscillation of the piston around the axis of the piston pin in the presence of combustion gases. Variations in the location of wear are the result of cold and hot spots caused by water circulation patterns around cylinders. Care should be taken to find the position of maximum wear beginning in a direction at a 90° angle to the piston pin and just below the ring ridge (see Fig. 6-1).

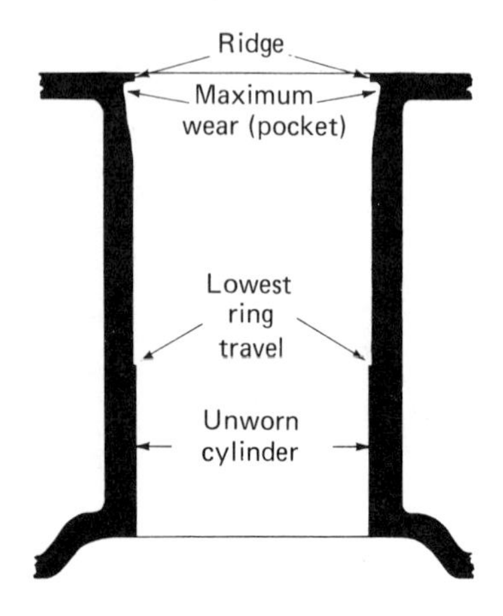


*FIGURE 6-1 The location of maximum cylinder wear (Courtesy of Sunnen Products Co.)*

The bottom portion of the cylinder below the piston ring travel will not wear to any measurable extent. The difference between the cylinder diameter at the bottom and the cylinder diameter at the most worn points at the top of the cylinder is called *taper.* Cylinder taper, and sometimes roundness, is used to determine whether cylinders are within service limits.

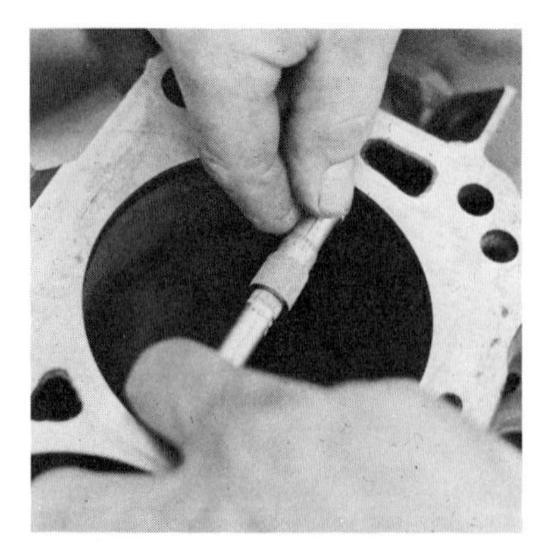

*FIGURE 6-2 Measuring a cylinder with an inside micrometer*

A common method of measuring cylinder diameter is with inside and outside micrometers. First, the inside micrometer is set to the cylinder size (see Fig. 6-2). Second, the diameter is determined by using an outside micrometer to measure across the inside micrometer. Inside micrometers may be read directly without an outside micrometer only if the tool is calibrated to the range of diameters being measured. This is often not the situation because it would require that inside micrometers be recalibrated each time extensions or tips are changed.

A second method of measuring is with a dial bore gauge (see Fig. 6-3). The dial of the gauge is set to read the specified cylinder diameter in a setting fixture or with a micrometer. The reading obtained in the cylinder will be the difference between the cylinder diameter and the diameter set in the setting fixture.

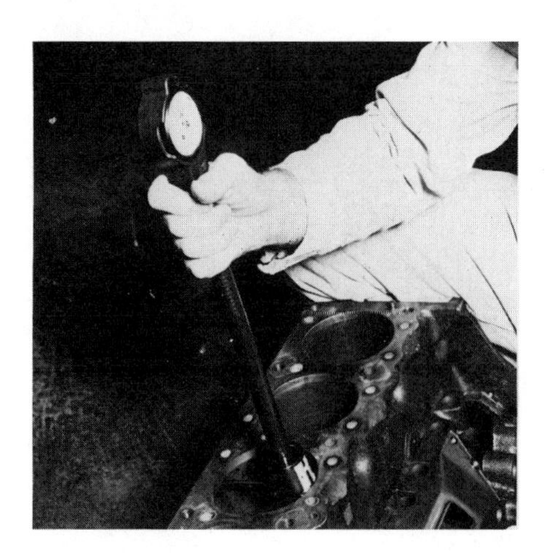

*FIGURE 6-3 Measuring a cylinder with a dial bore gauge*

It is recommended here that a service limit of .008 inch (.20 mm) of cylinder taper be used whenever manufacturers' limits are not known. One of the reasons for using this service limit is that cooling system variations and cylinder distortion frequently cause cylinder wear to occur at points not exactly opposite each other. In other words, cylinder wear cannot be accurately measured on the basis of diameters because measurements of diameter must be made in a straight line through the center of the cylinder, and the wear points may not be on that line. A set of piston rings for a standard bore diameter will often be marked "Std. - .010." Such rings are intended for use in standard bore diameter cylinders up to .010 inch (.25 mm) oversize. Considering that cylinder wear is likely to be greater than measured, the .008-inch (.20 mm) limit on cylinder taper is at least nearly suited to piston rings intended for use in cylinders up to .010 inch (.25 mm) oversize.

## MEASURING PISTON CLEARANCE

Piston clearance is not to be confused with cylinder taper. A cylinder with an .008-inch

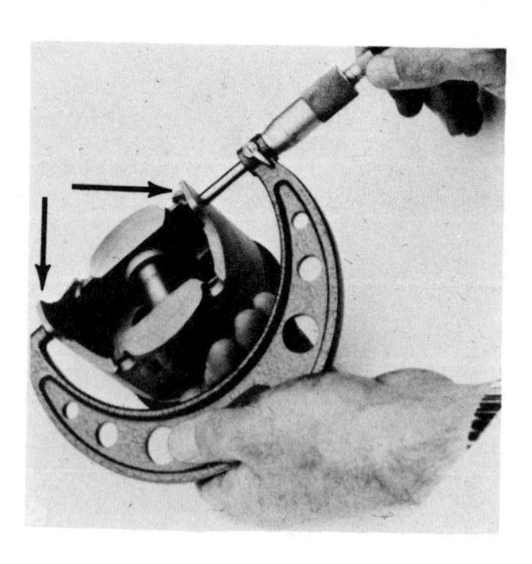

*FIGURE 6-4 Measuring piston diameter between the tips of the piston skirts and the piston pin centerline*

(.20 mm) taper does not necessarily have .008-inch (.20 mm) piston to cylinder wall clearance. Piston clearance is the difference between the minimum cylinder diameter and the maximum piston diameter. As mentioned in regard to cylinder taper, minimum cylinder diameter is found in the bottom of the cylinder below the ring travel. Maximum piston diameter is found by measuring midway between the piston pin centerline and the tips of the skirts perpendicular to the piston pin (see Fig. 6-4).

Engine and piston manufacturers make clearance recommendations for their products. A common serviceable range is .001 inch to .0025 inch (.02 mm to .06 mm). It is recommended that manufacturers' specifications be checked for the service limits on piston clearance. Piston clearance that is too loose will cause excessive noise, reduced piston ring life, and possible failure of the pistons.

## CHECKING PISTONS

There are three basic checks for the serviceability of pistons. The first one is for ring groove wear, especially on the top compression ring groove. The second check is for worn or collapsed piston skirts. The third is for cracked skirts or cracks around piston pin bores.

A commonly used service limit used for the ring groove is .006 inch (.15 mm). When the groove is checked with a gauge .006 inch (.15 mm) thicker than a new piston ring, the gauge should not enter the groove at any point (see Fig. 6-5). As mentioned, wear generally occurs on the top ring groove, where exposure to combustion gases is most severe. Worn ring grooves can be repaired by cutting them oversize and using a spacer to correct piston ring side clearance. Failure to correct this condi-

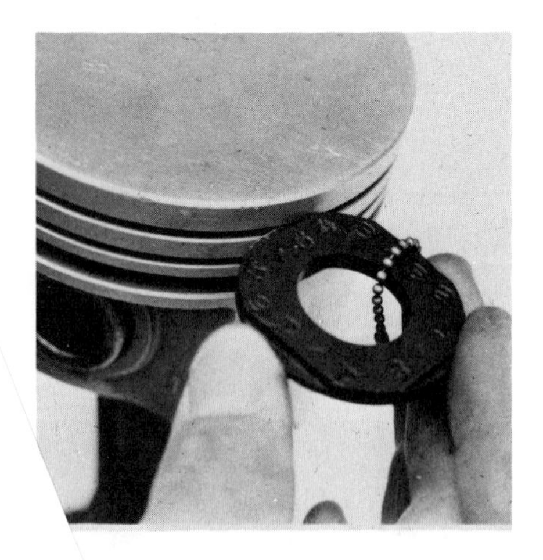

*FIGURE 6-5 Checking for ring groove wear with a Perfect Circle gauge*

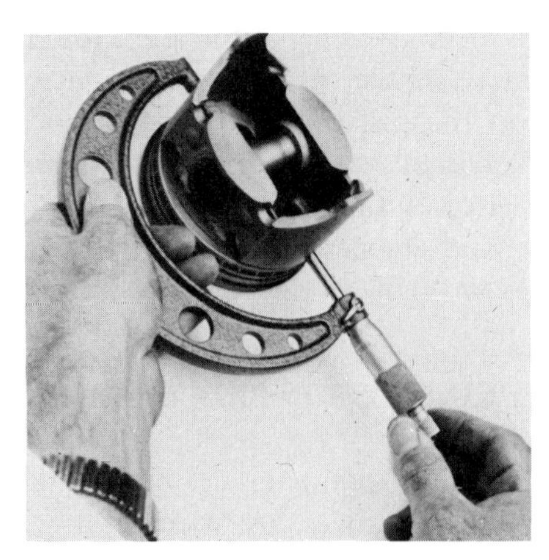

*FIGURE 6-6 Measuring piston diameter at the skirt near the piston crown*

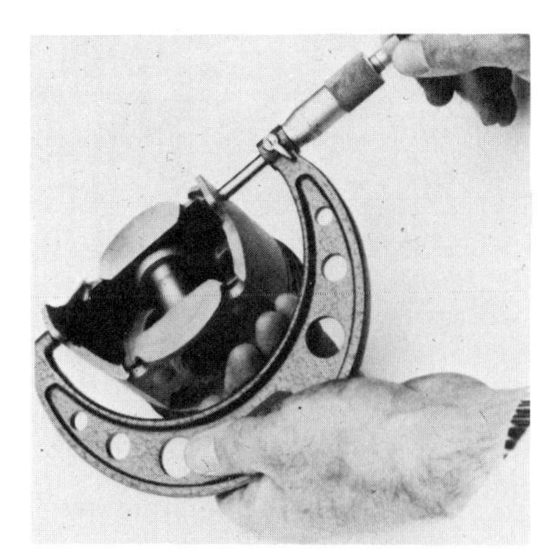

*FIGURE 6-7 Checking for piston skirt distortion by making a second measurement near the tips of the skirts*

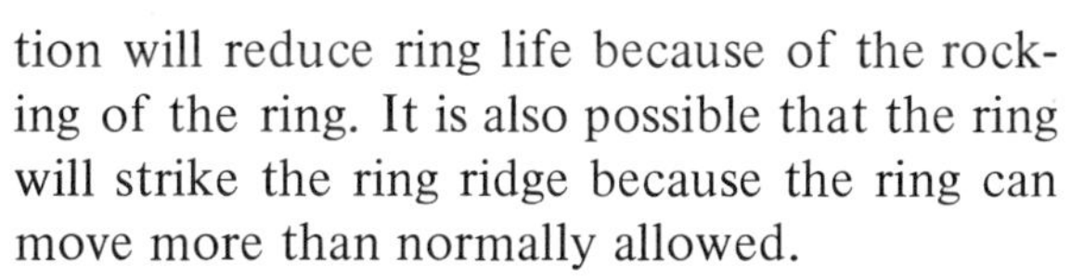

tion will reduce ring life because of the rocking of the ring. It is also possible that the ring will strike the ring ridge because the ring can move more than normally allowed.

Pistons must be checked for wear or distortion at the skirts by measuring at two points on the piston skirts (see Figs. 6-6 and 6-7) perpendicular to the piston pin. The diameter near the ends of the skirts should be equal to or up to .001 inch (.03 mm) greater than the diameter just under the ring lands. Do not measure across the ring lands because the piston will be .020 inch to .030 inch (.50 mm to .75 mm) undersize at that point. Pistons that measure up to .002 inch (.05 mm) undersize near the tips of the skirts may be expanded by knurling to make sure the engine runs quietly.

A new piston has a cam configuration such that the diameter is greatest perpendicular to the piston pin (see Fig. 6-8). When a piston is overheated or stressed, the cam shape may be lost or reduced. A piston in this condition is said to be "collapsed." Measurements

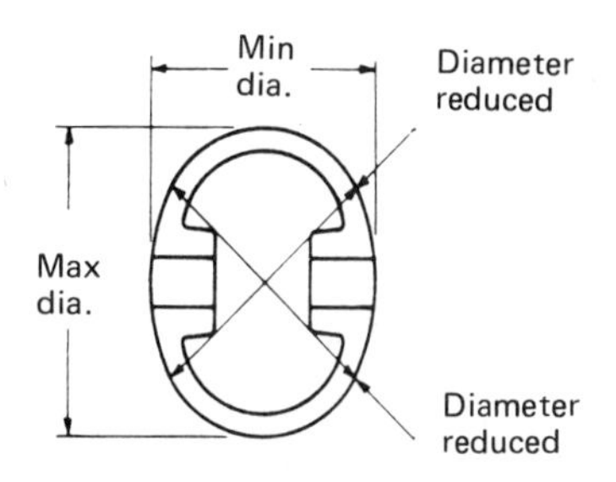


*FIGURE 6-8 The directions for measuring the piston cam configuration*

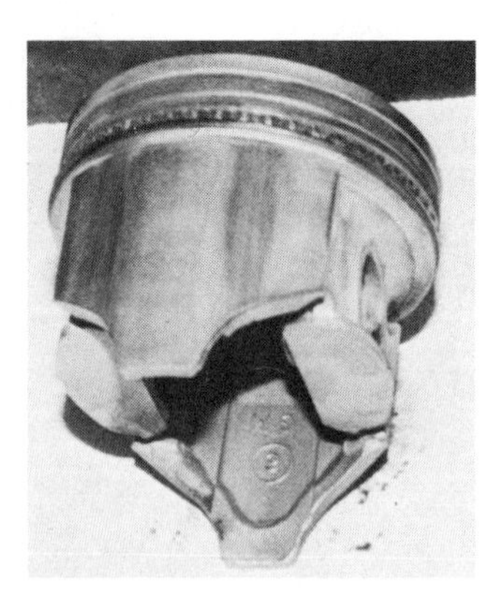

*FIGURE 6-9 The wear pattern associated with a "collapsed" piston*

taken at 45° to the piston pin should be considerably smaller than measurements taken at 90° to the piston pin. There will often be a characteristic wear pattern on the piston skirts associated with a collapsed piston (see Fig. 6-9). The engines will also have loud piston noises or "slap."

## CHECKING PISTON PIN CLEARANCES

For engine overhaul, it is desirable to determine whether clearances are within service limits. Checking these clearances does present special problems, however. For example, in oscillating piston pin designs, it is impractical to disassemble the piston from the rod. Efforts at disassembly and reassembly will almost certainly result in distortion of the piston. Because of the risk, clearances are frequently not measured during overhaul. Clearances can be checked by "feel," and obviously loose fits can be corrected by replacing worn piston pins or by fitting oversize piston pins.

Clearances for full-floating piston pins are much easier to check. The pistons are easily removed from the connecting rods, and measurements can be made directly. Clearances are determined by comparing the piston pin diameter to the bore diameters in the piston and in the connecting rod. Limits for clearances are specified by the manufacturers and are generally very small (approximately .0004 inch or .01 mm). Because of the precision, special gauging equipment is recommended for these measurements (see Fig. 6-10).

*FIGURE 6-10 Using a Sunnen AG300 gauge to measure piston pin bores*

## CHECKING CYLINDER BLOCK FLATNESS

It is occasionally found that the cylinder head gasket surfaces of an engine block will become warped. These surfaces should be checked for flatness to ensure head gasket sealing, especially if the engine has a history of overheating or head gasket failure.

The most common method of checking flatness is with a straightedge and a feeler gauge (see Fig. 6-11). A limit commonly used is .004 inch (.10 mm). That is, if a .004-inch (.10 mm) thickness gauge can be fitted between the straightedge and the block surface in any location, the block should be resurfaced. For this check, the straightedge is positioned diagonally across each pair of corners of the block and then across the cylinders. Care must be taken to clean and deburr the block surface prior to this procedure, or the results will be inaccurate.

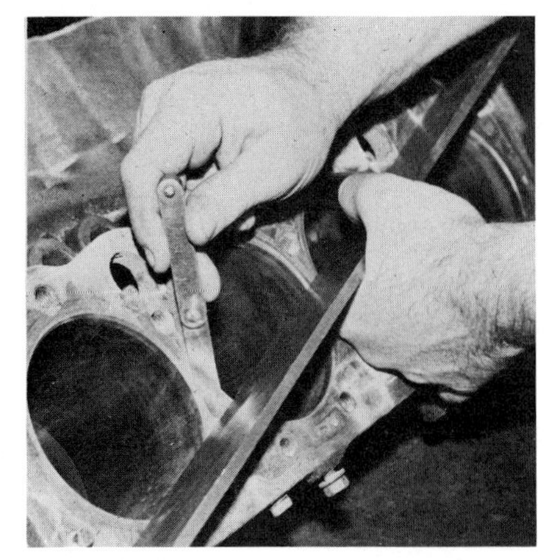

*FIGURE 6-11 Checking block flatness*

## MEASURING MAIN BEARING BORES

It is not unusual to find main bearing housing bores in engines elongated or stretched beyond the limits of bore diameter specifications. The stretching occurs primarily in the main bearing caps and is caused by crankshaft loads during severe operation. The problem is very apparent in the newer lightweight engine blocks used for high horsepower engines.

The preferred method for measuring main bearing bore diameters is to use a dial bore gauge (see Fig. 6-12) because of the close range of tolerance limits specified for bore diameters (typically .0005 inch to .001 inch or .013 mm to .025 mm). An inside micrometer or telescoping gauge may also be used, but the accuracy of measurements by these methods is entirely dependent on the mechanic's skill. Regardless of the tool selected, this check should not be omitted. Elongated bearing bores are a frequent problem. The faces of the block and the bearing caps should be cleaned carefully and deburred before

*FIGURE 6-12 Using a Sunnen dial bore gauge to check main bearing housing bores*

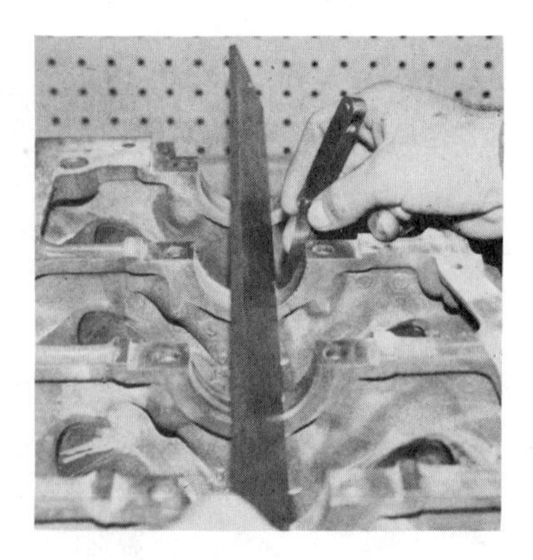

*FIGURE 6-13 Checking for warp along the main bearing housing bores*

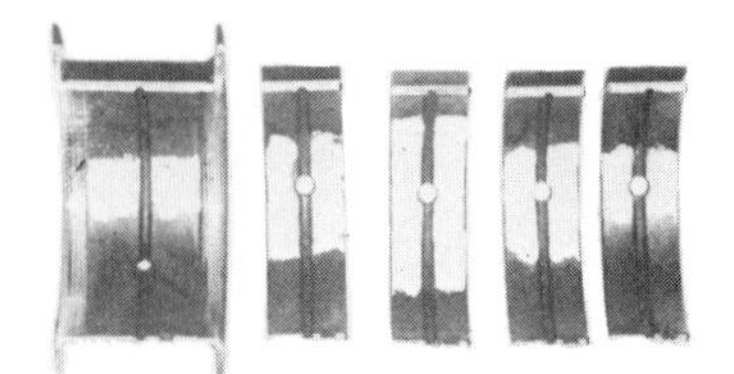

*FIGURE 6-14 Main bearing wear patterns from a bent crankshaft*

*FIGURE 6-15 Measuring rod journal near the fillet when checking for taper*

torqueing the bearing caps in position for these measurements.

Failure to detect discrepancies in bore diameters, and to correct them, may make it impossible to obtain main bearing oil clearances within the optimum range or even within specified limits. The fit of the bearings in the bores will also be incorrect.

Also check for misalignment along the main bearing bore center line. If severe enough, such misalignments are detectable with the usual straight edge and thickness gauge (see Fig. 6-13). Another method of detecting misalignment is to examine the wear patterns of the old main bearings. Bearing bore misalignment or a bent crankshaft can be seen in the bearing wear (see Fig. 6-14). Another method is to test for binding of the crankshaft in the bearings when the engine is assembled.

## CHECKING THE CRANKSHAFT

The crankshaft should be inspected for scoring, taper, or out-of-roundness. Scoring is usually a direct result of dirt in the oil. Journals, rod or main, develop a taper or out-of-roundness as a result of extremely high mileage or abuse.

Scoring of the crankshaft journals is generally evaluated visually. Extreme cases present no particular problem in making a judgment. It is in cases of minor scoring that judgment and experience are needed to decide on using the crankshaft or regrinding it. If it is expected that crankshaft polishing will not remove the scoring, regrinding is recommended.

Taper is measured with an outside micrometer (see Fig. 6-15). The journal diameter is measured at each outside edge near the *fillets* (the radius at each corner of a journal). The differences in the diameters is the taper. Recommendations vary on acceptable limits for taper. Considering the tolerances of most currently produced engines, it is recommended that a limit of .0005 inch (.01 mm) be used as a general guideline.

Out-of-roundness is also measured with an outside micrometer. The diameter of the journal is measured at three or more points around the circumference. The range of differences in these measurements is the amount of out-of-roundness. As with taper, a limit of .0005 in (.01 mm) is recommended as a basic guideline for determining whether regrinding is required.

The journal diameter must be within the specified limits. Keep in mind that the crankshaft being measured may have been reground before. If so, the diameter will be an even .010 inch, .020 inch, .030 inch (.25 mm, .50 mm, .75 mm) or more below the specified diameter.

The crankshaft should be checked for straightness. This may be done in V-blocks (see Fig. 6-16) or in the engine block with only the *end main bearings* in place (see Fig. 6-17). The dial indicator should not read greater than half of the specified main bearing oil clearance. The total indicator reading will be a combination of shaft roundness and straightness.

A method used to inspect for cracks where magnaflux inspection is not available is referred to as "ringing" the crankshaft. It is done by rapping the end of the crankshaft. If cracked badly, the shaft will not resonate–it will not "ring." All sprockets and keys must be removed for testing. The test does not disclose all flaws, and magnaflux inspection is still recommended, especially if the crankshaft is to be used in a heavy-duty application. Such magnaflux procedures are covered in the next part, "Crack Detection."

A crankshaft suitable for engine overhaul is straight, has a good surface finish and a taper and out-of-roundness on journals limited to .0005 inch (.013 mm). The diameter must also be within specified limits or an even

*FIGURE 6-16 Checking for a bent crankshaft in V-blocks*

*FIGURE 6-17 Checking crankshaft straightness in an engine block with only the end bearings in place*

amount undersize because of regrinding. It will be found that many engines, with reasonable care, will have serviceable crankshafts after 100,000 miles or more.

## MEASURING CONNECTING ROD BORES

The connecting rods of many engines that are currently manufactured are of a lightweight design. The advantage to keeping the weight down on connecting rod and piston assemblies is reduced inertia. The effect is a reduction or lowering of the forces that act on the connecting rod bearings each time the direction of travel is reversed. This is especially important in high-speed engines.

It is common to find that the housing bores of connecting rods become stretched or elongated. The amount of bore stretch varies, but it is not unusual to find diameters as much as .002 inch to .003 inch (.05 mm to .08 mm) over specifications. Years ago mechanics probably found that checking for bore stretch was unnecessary. Today, not checking for bore stretch in our newer engines could well lead to rod or bearing failure.

*FIGURE 6-18 Gauging connecting rod housing bores on a Sunnen AG300 gauge*

There are special tools available for checking the connecting rod bores (see Fig. 6-18). The out-of-roundness limit for engine overhaul is .001 inch (.025 mm), and diameters should be within specified limits. These bores can be measured with an inside micrometer, but keep in mind that the accuracy of measurements will be limited.

# 7

# Crack Detection

Before investing the time and effort in reconditioning cylinder heads and blocks, it is wise to check for cracks. Cracks may be caused by extremely high or low temperatures or too rapid a change in temperature. Many cracked castings may be salvaged at a considerable savings over replacement costs.

Connecting rods and crankshafts also develop cracks that can cause engine failure. It is true that these parts in passenger car engines are often not inspected for cracks. However, inspection will reveal cracks that have not caused failure *yet,* but the risk in reusing them is a little frightening. Inspecting connecting rods and crankshafts in racing and heavy-duty engines is a must. Cracked connecting rods and crankshafts are replaced.

*FIGURE 7-1 A crack in a cylinder head made visible by magnetic particle testing*

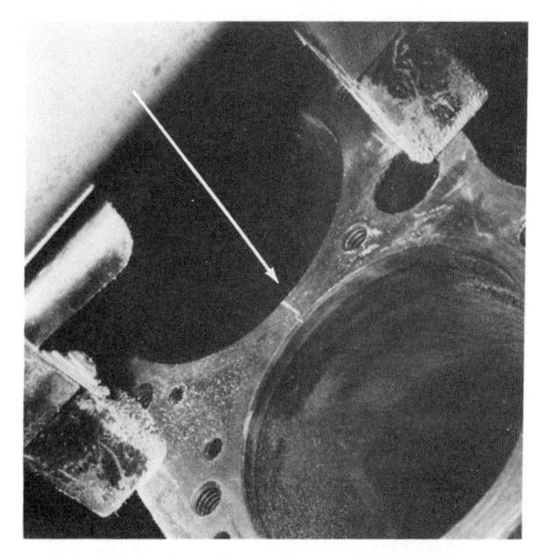

*FIGURE 7-2 A crack in a cylinder block made visible by magnetic particles*

## MAGNETIC PARTICLE INSPECTION

Magnetic particle inspection can be used with iron and steel engine parts. An intense magnetic field is set up in the part being tested. The area is then dusted with magnetic powder. Interruptions in the magnetic field due to a crack cause magnetic lines of force to form outside the part. The magnetic powder collects at the lines of forces paralleling the crack (see Figs. 7-1 and 7-2). Because the magnetic field is directional between the magnetic poles, it is necessary to test across the part in different directions.

Magnetic particle testing as described is used for cylinder heads and engine blocks. However, the crack must be across a surface which is visible. Common areas checked by this method include combustion chambers and ports, core holes, block surfaces, and main bearing webs. Internal cracks in castings cannot be detected by this method because they are not visible from the outside.

*FIGURE 7-3 Spraying a crankshaft with a solution carrying the magnetic particles*

## WET MAGNAFLUX

Wet magnaflux methods are used to detect cracks in connecting rods and crankshafts. The operation also involves magnetic particles, but the process is more sensitive. The magnetic particles are fluorescent and are suspended in oil or water. The solution is applied to the part by dipping or spraying (see Fig. 7-3). The part is then placed in a magnetic field and viewed under a black light (see Fig. 7-4).

This process is considerably more sensitive than dry testing in normal light. In fact, it will be found that virtually all forged parts

*FIGURE 7-4 A crankshaft viewed under a black light*

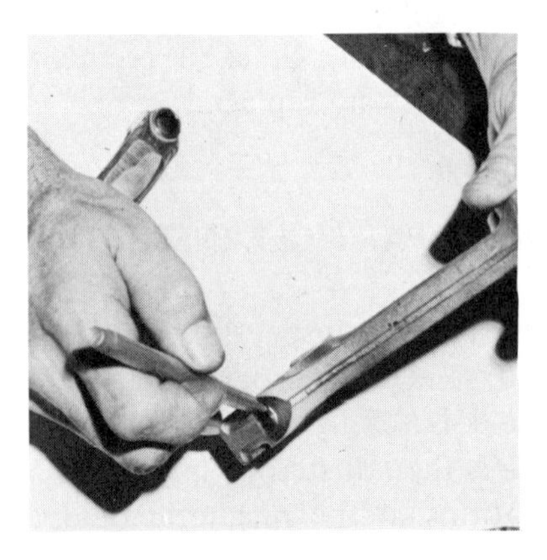

*FIGURE 7-5 One location for cracks in connecting rods*

will have some cracking evident. Many of these cracks are superficial only and do not lead to failure. Experience is required to determine whether parts are serviceable or should be rejected. Some examples of cracks that are absolutely unacceptable are cracks around rod bolt holes (see Fig. 7-5) and around the fillets on crankpins (see Figs. 7-6 and 7-7).

*FIGURE 7-6 Watch for cracks in fillets around crankpins*

## DYE PENETRANTS

Testing by the dye penetrant method may be used on all materials whether or not they are magnetic. As a result, the process is used in the automotive industry for testing aluminum cylinder heads and engine blocks.

The parts must first be thoroughly degreased and decarbonized. Penetrant is applied to the parts by dipping, spraying, or brushing (see Fig. 7-8). A few minutes are allowed for the penetrant to enter any pores or cracks. Excess penetrant on the surface is then washed off with a remover, and the surface is rinsed

*FIGURE 7-7 Crankshaft failure begun with a crack in the crankpin*

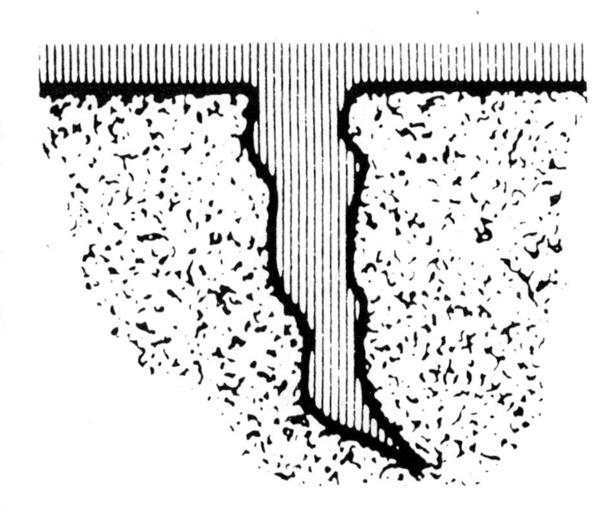

*FIGURE 7-8 Dye penetrant applied to the surface*

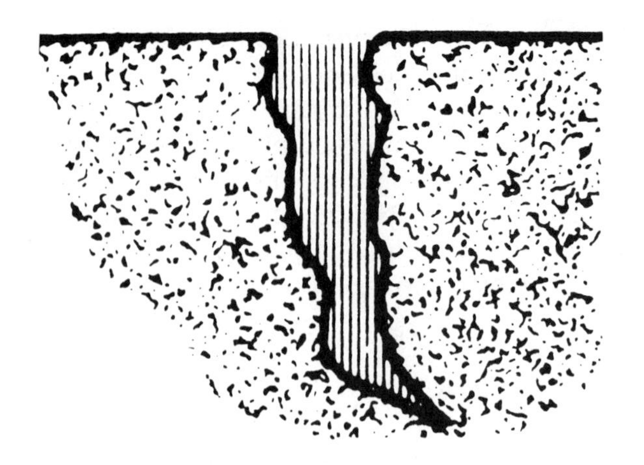

*FIGURE 7-9 Dye penetrant removed from the surface*

with water and wiped dry (see Fig. 7-9). Developer is then sprayed over the surface. The developer dries to a powdery film that draws the penetrant out of cracks. The cracks appear as lines through the developer (see Fig. 7-10). This process is made more sensitive by using a flourescent penetrant and a black light as in the wet magnaflux method.

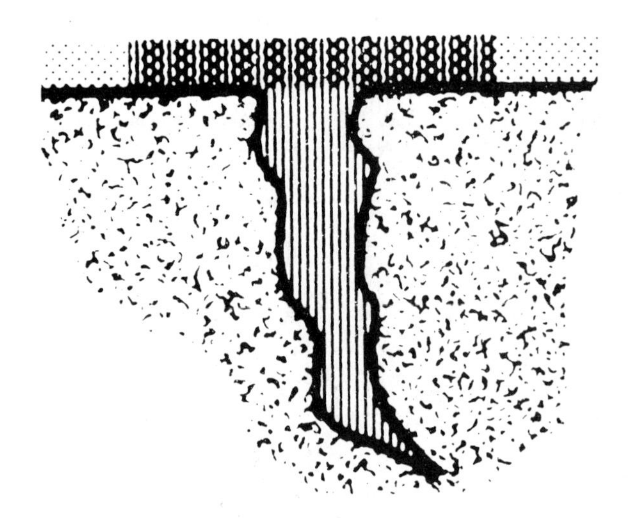

*FIGURE 7-10 Dye penetrant showing a crack after developer is applied to the surface*

## PRESSURE TESTING

The most effective means of finding cracks in iron or aluminum cylinder heads and engine blocks is by pressure testing. Water passages in the part are blocked off, and the water jacket is pressurized up to 55 psi (see Fig. 7-11). Then the surface of the casting is sprayed with a solution that bubbles if air leaks through a crack. This process is perhaps the only way to locate cracks in the oil galleys, through oil return holes, or in lifter bores and other areas that are not visible from the outside.

*FIGURE 7-11 An Iron-Tite pressure tester with cylinder head in place and air being connected*

# 8

# Crack Repair

As mentioned earlier, many cracked cylinder heads and blocks can be salvaged with relatively inexpensive repair procedures. Many of these procedures have been developed over the years to repair costly industrial and diesel engine castings. Some of these procedures can be economically applied to repairing automotive castings.

## USING THREADED TAPER PINS

Threaded taper pins may be used to repair cracks on the surfaces of cylinder heads and blocks. The limitation is that it must be possible to reach the full length of the crack with a drill, a reamer, and a tap used in making the repair.

The pins are a high quality ductile iron material for repairing iron castings. Aluminum pins are available for repairing aluminum castings. The pins are available in different sizes to suit the job (see Fig. 8-1). The procedure basically involves drilling, tapping, and installing pins along the length of the crack.

The first step is to locate *exactly* where the crack begins and ends. Typically, this means using a magnetic particle crack detection process. Of course, dye penetrants may

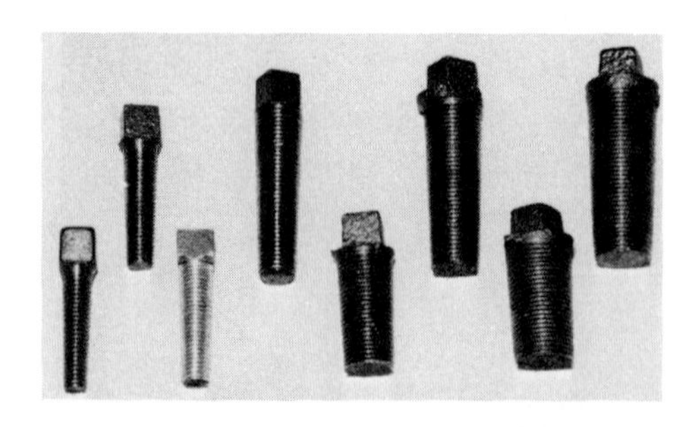

*FIGURE 8-1 Assorted threaded taper pins indicating the range of available sizes*

be used for aluminum castings. This step is critical to make sure that the full length of the crack is repaired.

The next step is to drill a hole 1/8 inch (3 mm) *past* each end of the crack (see Fig. 8-2). This stops the progress of the crack during repair. It is essential that the hole be drilled past the end of the crack as seen on the outside of the casting because the crack generally extends farther on the interior of the casting. The drill size is the same as required for tapping threads in the hole. Both drill and tap sizes will depend on the size of the pin used. Check again with magnaflux or penetrant to be sure the crack does not continue past the drilled holes.

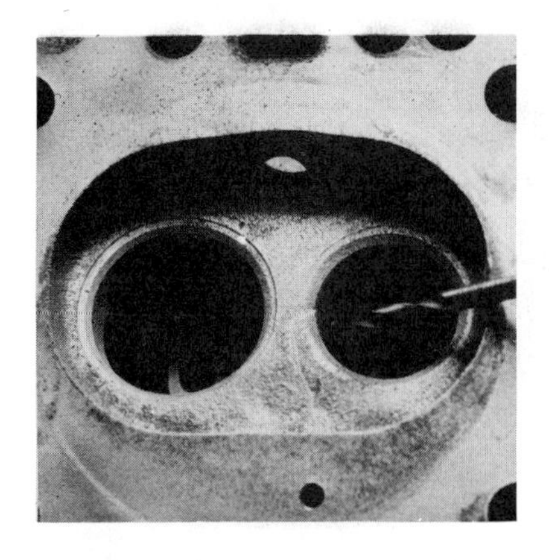

*FIGURE 8-2 Drilling holes past each end of the crack*

The drilled hole at one end is then tapped for the threads on the pin (see Fig. 8-3). It is recommended that holes drilled 1/4 inch (6 mm) deep or more be reamed with a tapered reamer to make tapping the holes easier. Be careful not to tap too deeply. The tapered pin may go too far through the hole before the threads tighten. The tapered pin is then coated with ceramic sealer and tightened into the tapped hole.

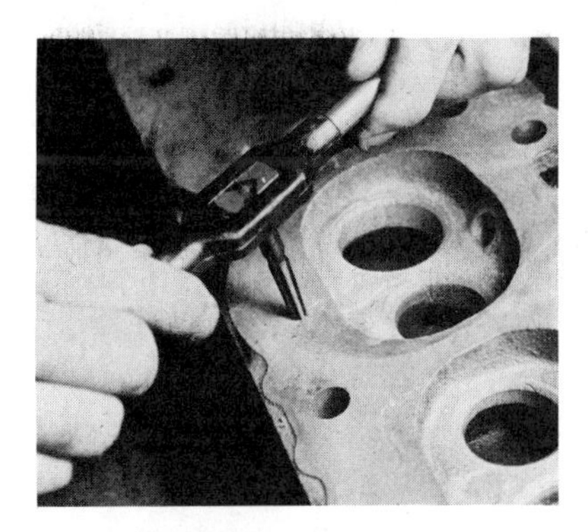

*FIGURE 8-3 Tapping the hole for a threaded pin*

The excess length of pin must be removed. One method is to saw part way through the pin at about 1/16 inch (1.5 mm) above the surface of the part (see Fig. 8-4). The pin is

*FIGURE 8-4 Sawing off the excess length of pins*

*FIGURE 8-5 Peening the threaded pin and the surrounding casting*

then snapped off. Because the pin may break off below the surface, one recommendation is that the pin be sawed all the way through if possible. Another trick is to use a drill to remove the pin by drilling along the length protruding out of the work. Once the pin is drilled, a sharp chisel will easily shear the pin through. This method is especially effective when pins cannot be easily reached with a saw.

The end of the pin is now peened (see Fig. 8-5) to expand the pin outward into the work. The work may also be peened inward on the pin. The process of installing pins is then continued along the length of the crack.

The pins may be made to seal the crack by two basic methods. The first is to overlap the pins. After installing the pin, the next hole is drilled so that the edge of the hole just contacts the edge of the pin. The result is that after tapping the hole the next pin will overlap the first. In this way, the crack is sealed and the pins are locked in place by each other (see Fig. 8-6).

The second method is to place the pins so that they intersect (see Fig. 8-7). In this method, parts of the crack will remain visible

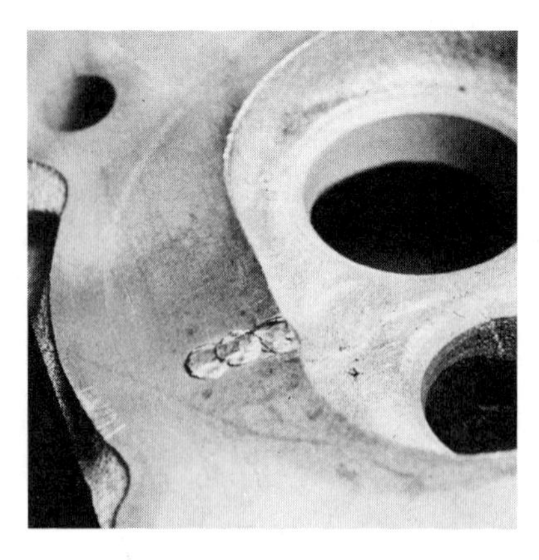

*FIGURE 8-6 The overlapping of threaded pins as a method of repair*

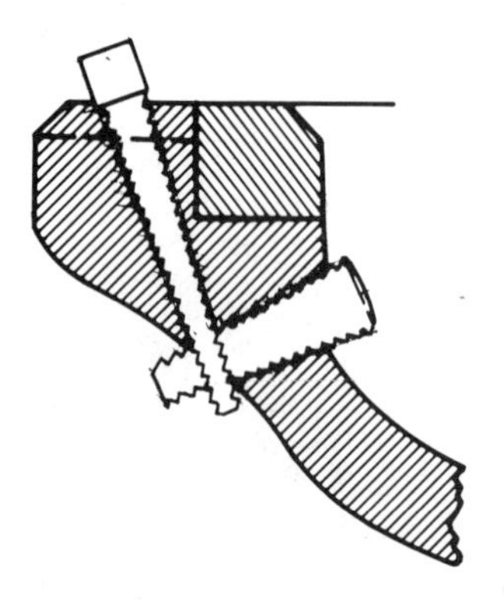

*FIGURE 8-7 The intersecting of threaded pins as another method of sealing a crack*

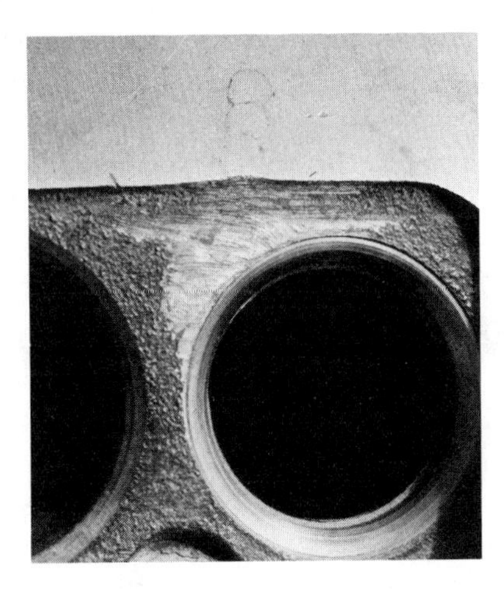

*FIGURE 8-8 A valve seat inserted as a part of crack repair*

*FIGURE 8-9 Grinding the combustion chamber surface after installing threaded pins to repair a crack*

on the outer surface, but the crack will be sealed below the surface.

Cracks frequently extend across valve seats. The usual method of repair in this case is to use overlapping pins starting at each end and working toward the seat. A valve seat insert is then installed (see Fig. 8-8). The insert locks the pins in place and ensures a good surface for the valve seat.

The final part of the job is to blend the repaired surface with the original surface. This is done with grinding stones or rotary files (see Fig. 8-9). The flat surfaces of blocks and heads may be draw-filed, although it may be more convenient and cleaner to perform resurfacing operations last and to clean up crack repairs in the process.

A ceramic sealer is used to follow up crack repairs to make sure that even the smallest of leaks is sealed. For example, when a repaired cylinder head is installed on an engine, ceramic sealer is added to the cooling system. Exact procedures usually call for cleaning the system before adding the sealer.

The sealer is then run in the system for five to seven days. The coolant is then drained, and the sealer left to cure for three hours. After this treatment, the water jacket is lined with ceramic sealer.

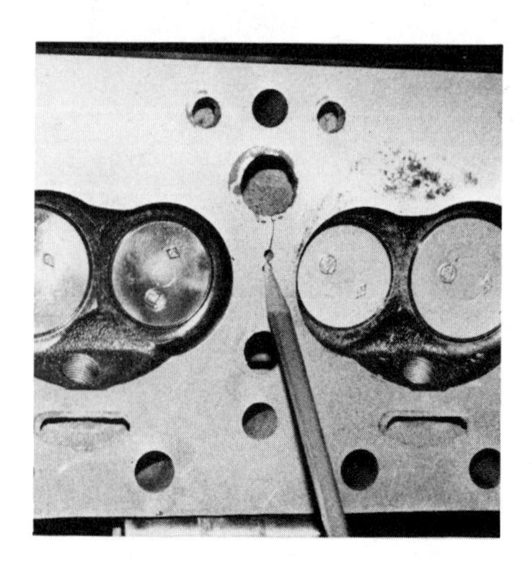

*FIGURE 8-10 An example of stop drilling to prevent the extension of a crack*

## STOP DRILLING

Some cracks will not cause problems if left unrepaired. These are cracks that do not cross oil passages, bolt holes, or seal surfaces. It is a good idea, however, to make sure the crack does not grow larger. The growth of the crack may be stopped by drilling a hole about 1/8 inch (3 mm) in diameter just past the end of the crack. A crack that starts outward from a water circulation hole in a cylinder head or block is an example of a crack that can be repaired with stop drilling (see Fig. 8-10).

## CERAMIC SEAL CIRCULATORS

*FIGURE 8-11 Circulator being connected to a cylinder head*

The procedure for adding ceramic sealer to a cooling system has already been covered. However, the ceramic sealer treatment is used in machine shop facilities on cylinder heads and blocks prior to assembly. The process requires the pressure tester as used for crack detection and a circulation system for the ceramic sealer solution.

The circulator heats the solution and pumps it through the casting under pressure. The cylinder head or block is mounted in the pressure tester, and water passages are blocked off. The circulator is then attached to the casting, and the solution is pumped through the water jackets for about fifteen minutes (see Figs. 8-11 and 8-12). Circulator inlet and outlet hoses are then shut, and one of the hoses is disconnected. The pressure tester is connected in place of the circulator, and the casting is momentarily pressurized (55 psi). The momentary high pressure packs the ceramic sealer into any separations in the casting. The casting is then drained and left to cure for at least thirty minutes.

This repair is successfully used as the complete repair procedure for some cracks. It is ideal for cracks that develop internally through the water jackets and across oil passages. Such cracks are, of course, inaccessible for repair with tapered pins.

*FIGURE 8-12 An Iron-Tite circulator*

# 9

# Reconditioning Valve Train Components

A discussion of procedures for performing valve service operations is complicated by the variety of tools and equipment and by the number of methods for each operation. The recommendations made here will cover some of the more prevalent practices.

The most important thing to consider is the goal in performing this phase of engine work. A "valve job" as discussed here is expected to provide proper valve sealing and longevity of service comparable to original equipment. It is also understood that on-the-job compromises are often made to keep service within customer budget limitations.

## REMOVING AND REPLACING VALVE GUIDES

Many engines come with replaceable valve guide bushings. These are perhaps the easiest to repair because worn guides may be driven out and new guides driven in.

There are some precautions, however. Some guides may be driven out in only one direction. Although this is usually from the valve port side, manufacturers' service references should be checked first. It is occasionally found that removing the valve guide "broaches" the hole, leaving it oversize. Oversize valve guides are sometimes provided by the manufacturer in the event that this occurs.

One method of making removal easier is to tap one end of the valve guide and screw in a bolt. A punch may then be used to drive out the guide by driving against the bolt rather than the guide (see Fig. 9-1). This method is especially helpful on copper or bronze guides, which deform easily when drivers are used directly against the shoulder of the guide.

*FIGURE 9-1 A bolt threaded into a valve guide. A driver is then used against the bolt.*

Precautions should be observed on installation of the new guides. First, measure the guide bushing and bore diameters to be sure the fit is correct. Be sure to lubricate the new guide bushing and the bore to prevent galling or broaching on installation. The guide should be installed to the height specified (see Fig. 9-2).

A couple of ways of making this job easier are used by some machinists. One is to press guides in rather than drive them in with a hammer and driver. This is not always possible without additional tooling because of the angle of the valve guides through the cylinder head. Another trick is to heat the cylinder head casting to 200°F to 300°F to minimize

*FIGURE 9-2 Checking the height of a valve guide above the cylinder head*

*FIGURE 9-3 Hand reaming a valve guide*

*FIGURE 9-4 A close-up view of a knurling arbor. Note the pilot and that there are no cutting edges*

the interference fit during installation. This last trick is especially helpful on aluminum cylinder heads.

It is recommended that a valve guide reamer be run through each valve guide after installation (see Fig. 9-3). This is in case the guides are deformed at the ends from the use of a driver for installation.

## VALVE GUIDE KNURLING

A commonly used procedure for resizing worn valve guides is *knurling.* This procedure calls for first running a "cold forming" tap with a pilot—the tap is called a *knurling arbor*—through each valve guide (see Fig. 9-4). The knurling arbor produces a thread in the guide by deforming the metal and forcing the crest of the thread inward, making the inside diameter of the valve guide smaller. It is important to note that the thread is *formed* and *not cut* as with a conventional tap. The knurling arbor is turned through the valve guide with a drill motor, speed reducer, and an extreme pressure lubricant (see Fig. 9-5).

Once knurled, the valve guide is reamed to provide clearance. A valve guide reamer is

*FIGURE 9-5 Knurling a valve guide using a drill motor and speed reducer*

special in that it has a pilot to keep the reamer in alignment (see Fig. 9-6). The size of the reamer used is approximately .001 inch to .002 inch (.025 mm to .050 mm) greater than the stem diameter. This size should produce a valve guide very nearly the same diameter as the original. It should be noted that minimum specified clearance may be used with knurled guides without concern for sticking valves because of the retention of lubricant in the groove or thread.

Reaming may be done by hand, but most production shops use a drill motor with a speed reducer (see Fig. 9-7). If hand reaming, be sure to turn clockwise only; reversing direction will dull the reamer. No cutting oil or lubricant is used with cast iron. Extreme pressure lubricant, or cutting fluid is recommended for bronze valve guides.

The knurling arbors commonly used for domestic passenger cars are for 5/16-inch, 11/32-inch, and 3/8-inch valve guides. They are also available .005 inch over standard sizes for use with oversize valve stems.

Care must be taken in selecting the correct tool for the guide diameter. A 3/8-inch knurling arbor will start in an 11/32-inch valve guide, but it will break before completing the operation. Check the size of the knurling arbor before using it. Dull knurling arbors will also readily break off in the valve guide.

Reamers are available in standard valve guide diameters for each application and in .001-inch (.025 mm) increments over the standard sizes to provide specified clearance for oversize valve stems. Metric sizes are also available, and it will be found that some English and metric sizes may be used interchangeably.

It is recommended that knurling not be used to correct wear exceeding .006 inch (.15 mm). In fact, many machinists use a wear

*FIGURE 9-6 A close-up view of a valve guide reamer. Note the pilot used to help align the reamer in the guide*

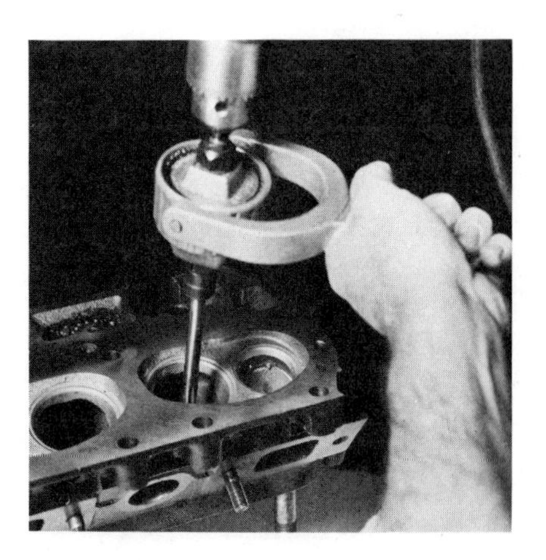

*FIGURE 9-7 Reaming is also done with a drill motor and speed reducer*

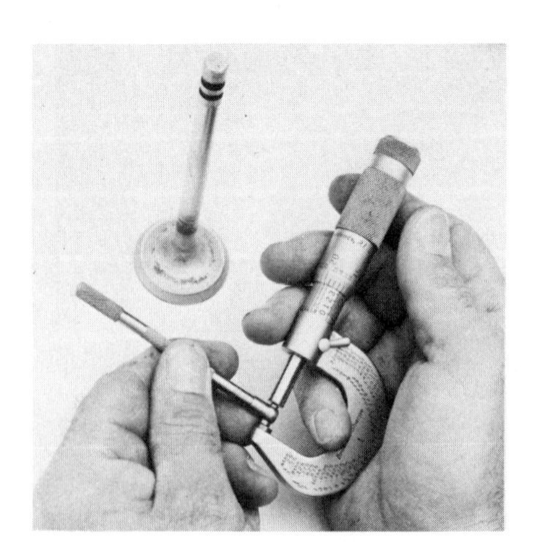

*FIGURE 9-8 Using a telescoping gauge set to the "no-go" limits of wear to determine whether knurling will be satisfactory*

limit of approximately .004 inch (.10 mm) if original equipment longevity is sought.

A quick method of determining whether wear is within limits for knurling is to set a telescoping gauge to the valve stem diameter plus .006 inch (.15 mm) (see Fig. 9-8). The telescoping gauge is then used as a "no-go" gauge for checking the valve guide wear. The valve guide is beyond limits for reconditioning by knurling if the gauge enters the valve guide in any position.

The finished diameter of a reamed valve guide should be measured before assembly because the reamers cannot be relied on to produce exactly the size indicated. Variations are generally caused by tool wear. Trusting the valve guide reamer to produce the diameter called for might well result in insufficient clearance and a stuck valve.

Knurled valve guides are especially difficult to measure because a groove or thread remains inside. For improved accuracy in measuring, a valve guide bore gauge is recommended for checking the finished job. The gauge probe is first set to the diameter of the valve stem (see Fig. 9-9). The indicator then reads the

*FIGURE 9-9 Setting the Sunnen P310 gauge to the valve stem diameter*

amount of clearance in the valve guide (see Fig. 9-10). This tool works especially well for knurled guides because the probe will span the internal groove or thread in the valve guide (see Fig. 9-11).

Be sure to scrub the valve guides clean after knurling and reaming. Chips are readily trapped in the thread left by knurling and reaming and can be removed only by a thorough scrubbing with cleaning solution and a bore brush. Failure to clean the valve guides will result in a stuck valve or rapid guide and stem wear.

Knurling is ideally suited to passenger car engines with "integral" valve guides—guides that are a part of the cylinder head casting and are nonremovable. Knurling cannot be used on the hardened valve guides used in some heavy-duty engines.

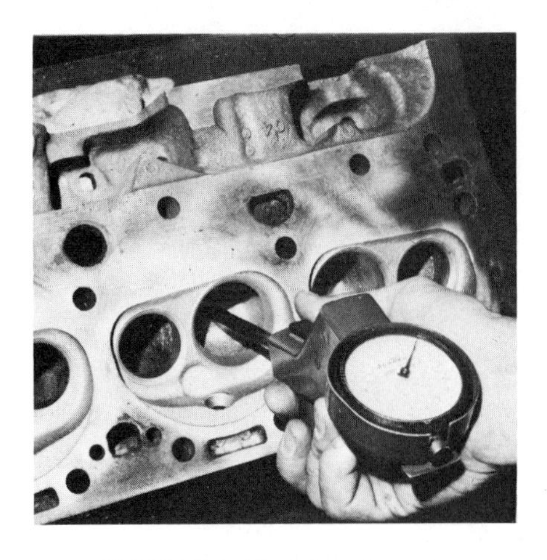

*FIGURE 9-10 Gauging the valve guide clearance*

*FIGURE 9-11 A close-up view of the expanding probe used in the Sunnen P310 gauge*

## FITTING OVERSIZE VALVE STEMS

Valves with oversize stems are available for many domestic engines as a means of correcting valve guide wear. The sizes typically available include .003 inch, .005 inch, .010 inch, and .015 inch (.075 mm to .380 mm) over standard size. They are especially suitable when valves must also be replaced. The valves with oversized stems can serve to correct guide wear and replace valves not within service limits. The valve guides are reamed for correct clearance with oversize stems. As mentioned earlier, reamers are available in the required sizes.

Keep in mind, however, that a valve guide reamer has a pilot to keep the reamer in alignment. Because of the pilot, each reamer can remove only about .003 inch to .005 inch (.075 mm to .125 mm). This could prove to be a problem when fitting oversize stems beyond the first oversizes because a wide selection of reamers is required so that reaming may be done in steps.

Valve seals may have to be changed when using oversize valve stems. Generally, rubber seals are acceptable for oversizes up to .015 inch and all Teflon® seals are limited to standard valve stem sizes.

The oversize valves, even the first oversizes, can be used in combination with guide knurling to correct severe valve and guide wear problems. There are other repair possibilities though, and the repair selected will depend on tooling on hand, parts available, and the customer's budget.

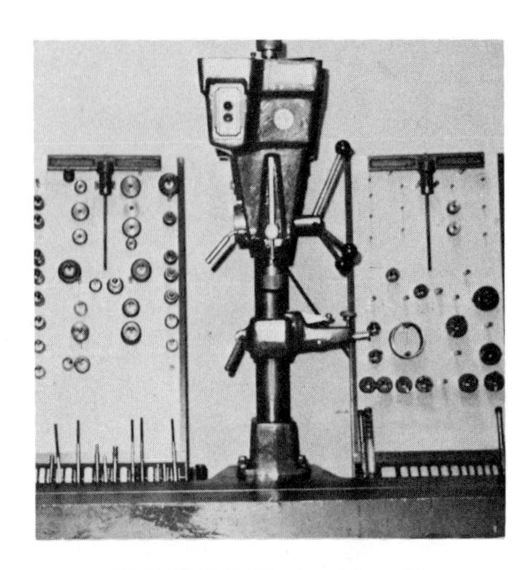

*FIGURE 9-12 A Van Norman IDL valve guide and seat machine, one of several machines available for cylinder head service*

## INSTALLING FALSE VALVE GUIDES

Cylinder heads with integral valve guides present special problems if extreme guide wear is present. As mentioned, valve guide knurling or oversize valve stems may be used to correct wear. However, when guide wear exceeds limits for knurling and valves are servicable, *false guides* may be installed in the cylinder head.

False guides are bushings similar to replaceable valve guides used in other cylinder heads. They are available in a number of sizes, the most common being for domestic passenger cars having 5/16-inch, 11/32-inch, and 3/8-inch (7.9 mm, 8.7 mm, and 9.5 mm) inside diameters. The guides for 5/16-inch valve stems commonly have 7/16-inch (11.11 mm) outside diameters. The 11/32-inch and 3/8-inch guides commonly have 1/2-inch (12.7 mm) outside diameters. They are available in cast iron or bronze.

*FIGURE 9-13 Core drilling through the worn valve guide in a Van Norman machine*

The guides are typically installed using machines and tooling designed for cylinder head service (see Fig. 9-12). The first step in machining is to drill the original guide approximately .025 inch (.635 mm) under the diameter of the replacement guide (see Fig. 9-13).

*FIGURE 9-14 Reaming to size for a false guide bushing in a Van Norman machine*

*FIGURE 9-15 A close-up view of the pilots on the core drill and reamer. The pilots and drill bushings align the cutting tools close to the original valve guide centerline.*

The next step is to ream the drilled hole to size (see Fig. 9-14). Note that drill bushings are used to align the cutting tools and that both the core drill and reamer have pilots (see Fig. 9-15). The drill bushings and pilots keep the centerline of the valve guide close to the original location.

The false guide is driven into place with a driver and hammer (see Fig. 9-16). The driver has a pilot that fits the inside diameter of the false guide. An extreme pressure lubricant should be used to prevent galling on installation. The guide is driven through until it hits an adjustable stop preset to the correct height (see Fig. 9-17).

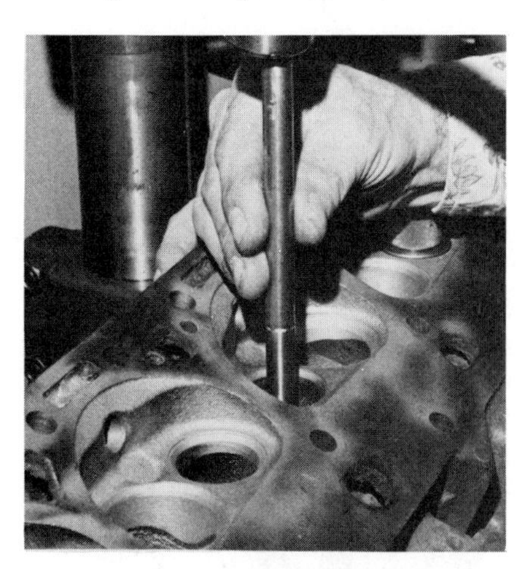

*FIGURE 9-16 Driving a false guide bushing. The driver must fit the guide bushing.*

*FIGURE 9-17 An adjustable stop set to the correct height for guide bushing installation*

*FIGURE 9-18 A cutoff tool being used to cut down a universal length valve guide bushing*

*FIGURE 9-19 The Van Norman cutoff tool and pilots*

False guides are typically stocked in universal lengths to minimize the inventory required. This means that there is frequently an extra length of guide extending into the valve port. The extra length is machined off using a cutoff tool (see Fig. 9-18). The cutoff tool can be fitted with different pilots according to the inside diameter of the guides (see Fig. 9-19). It is recommended that guides be deburred by reaming after the cutoff step.

False guides are especially good for engines that will be reconditioned repeatedly because the false guides can be removed and replaced during future service. If improved longevity is desired, valve guide materials can also be upgraded with bronze false guides.

*FIGURE 9-20 Refacing a valve in a Kwick-Way valve refacer*

## REFACING VALVES AND VALVE STEMS

Once it has been determined that valves are serviceable, they are refaced in specialized valve grinders (see Fig. 9-20). All pits must be removed in grinding because any that remain will cause hot spots on the valve face, and

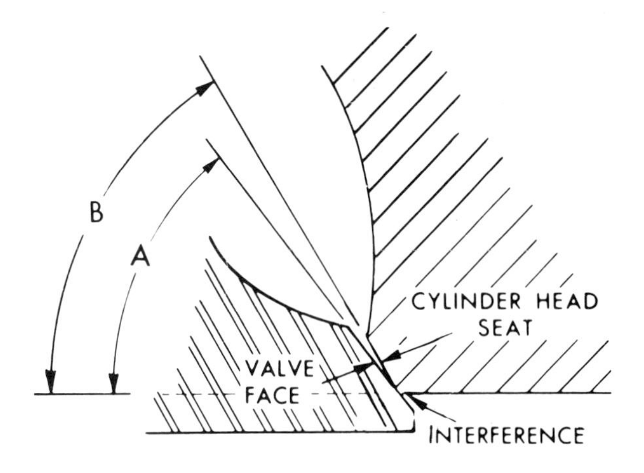


*FIGURE 9-21 The interference angle between the valve face and valve seat (Courtesy of Buick Motor Division)*

burning will result. The burning occurs because heat cannot transfer from the valve through the seat without contact at the pitted face through the seat at points where pitting remains. The surface finish quality is also as fine as can be obtained to ensure full contact on the valve seat. The grinding wheel must be grinding oil must be directed between the grinding wheel and the valve face to obtain the required surface quality.

Valve seat angles are approximately 45° or 30°; however, valves are typically refaced to provide a slight interference angle between the valve and seat (see Fig. 9-21). The interference angle provides for a narrow line of contact when the valve first contacts the seat. This angle tends to prevent the valve from trapping carbon under the valve face. The manufacturers often specify 45° (or 30°) for the valve face angle and a larger angle for the valve seat to provide a 1/2° to 2° interference angle.

Many machinists, however, use a 44° (or 29°) face angle and a 45° (or 30°) seat angle. This provides for 1° interference. One reason for this practice is that the slightly smaller face angle permits more grinding on the valve face without thinning the margin (see Fig. 9-22). More valves are salvaged in this manner,

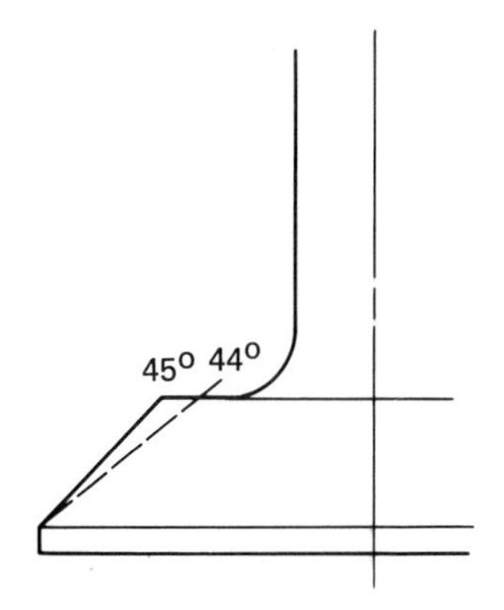


*FIGURE 9-22 Refacing to a slightly smaller valve face angle thins the margin the minimum amount*

and the slight variation from specifications does not reduce longevity or affect sealing.

It should be pointed out that warped or bent valves are easily detected during refacing. A warped valve will clean up on one side only (see Fig. 9-23). Discard these valves, and don't attempt to grind them to a cleanup.

The tips of valve stems must also be refaced. Check the chamfer (the bevel on the edge) first and regrind it if it is less than approximately 1/32 inch (.7 mm) wide (see Fig. 9-24). This amount will keep a sharp edge from forming after refacing the stem. Avoid chamfering stems *after* refacing because the grinding will raise burrs on the face that will have to be removed by hand. The stems are refaced at one end of the valve grinder (see Fig. 9-25). Refacing approximately .003 inch (.076 mm) will usually provide a smooth, flat surface for the rocker arm. Chamfering prevents a sharp edge from shaving the rocker arm face.

*FIGURE 9-23 A warped valve showing contact with the grinding wheel on one side only*

*FIGURE 9-24 Chamfering a valve stem on a Sioux valve grinder (Courtesy Sioux Tool Co.)*

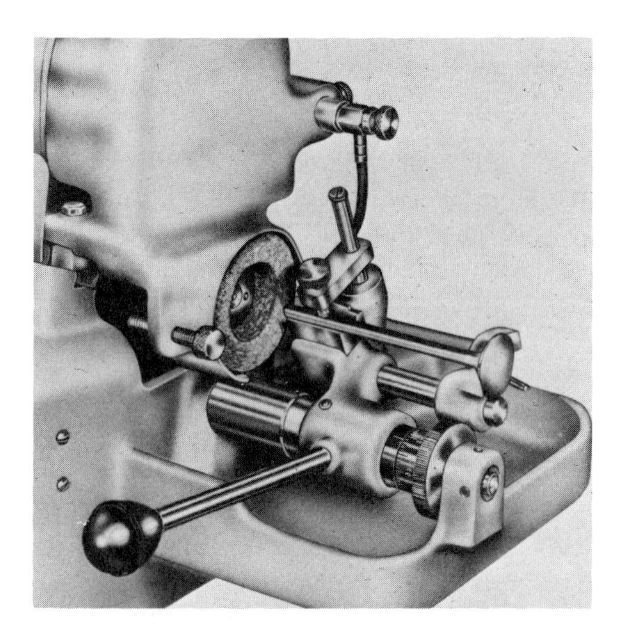

*FIGURE 9-25 Refacing a valve stem (Courtesy of Kwik-Way Co.)*

## GRINDING VALVE SEATS

It is absolutely essential that valve guides be in good condition before seat grinding because seat grinding stones are guided by a pilot in the valve guide (see Fig. 9-26). The pilot cannot properly locate in a badly worn valve guide. Incorrect location of the pilot will cause the valve seat to be ground off center or at an angle to the valve guide. A valve cannot close and seal under these conditions.

Valve seat inserts, if used, are generally checked at this time for visible cracks and for looseness. Looseness can be checked by placing a finger on the seat insert and rapping the insert lightly on the opposite side with a hammer (see Fig. 9-27). A loose insert will move when hit, and the motion can be felt through your finger on the opposite side. Loose or cracked seats must be replaced (see the next section, "Installing Valve Seat Inserts").

The object of valve seat grinding is to obtain a valve seat with the correct width and uniform width all around. It must also be located in the correct position on the valve face. High speed drive motors are used to obtain high quality valve seat finishes (see Fig. 9-28).

Valve seat width is critical. Too wide a valve seat tends to trap carbon and cause burning; too narrow a valve seat will not transfer sufficient heat and, therefore, causes burning. An average seat width is 1/16 inch (1.5 mm). Variations in seat width also cause differences in cooling and, therefore, possible distortion and subsequent burning.

The seat must contact the valve face slightly above the center of the face. Remember that seat grinding is being done at room temperature, but valves will be heated to 1400°F or more at operating temperature.

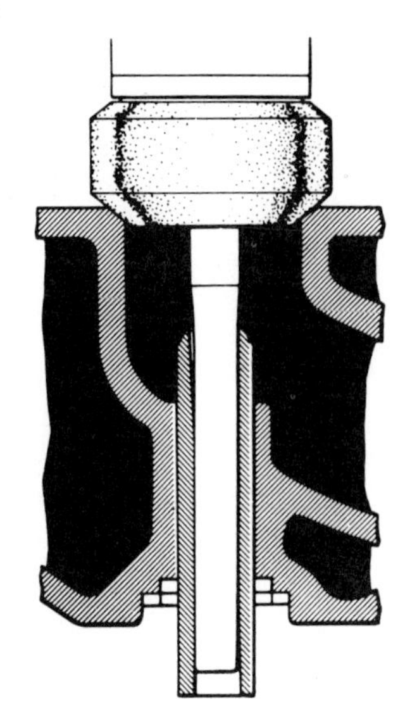

*FIGURE 9-26 The positioning of a valve seat stone over a pilot in the valve guide (Courtesy Sioux Tool Co.)*

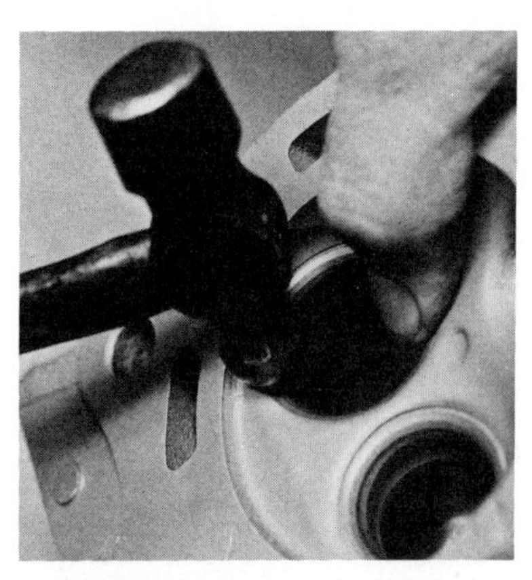

*FIGURE 9-27 Checking for a loose valve insert*

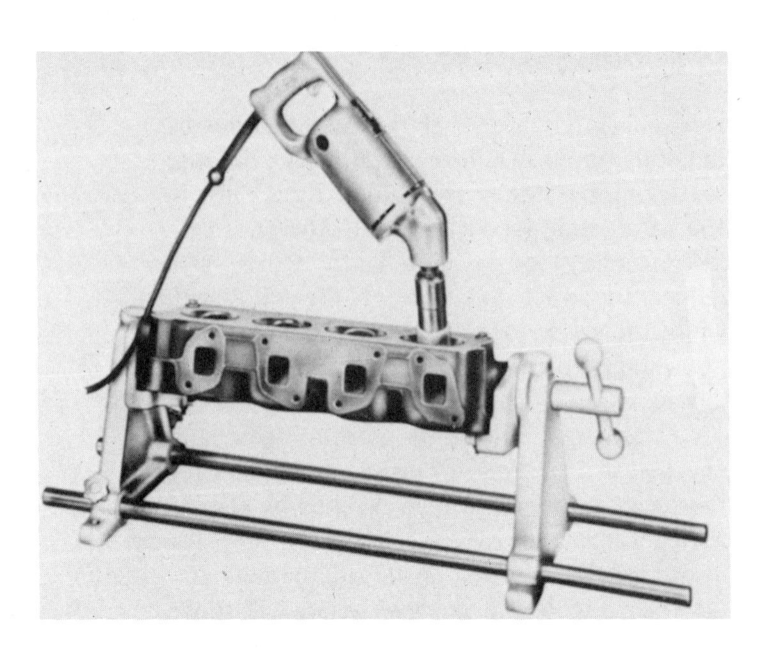

*FIGURE 9-28 A Kwik-Way hard seat grinder (Courtesy Kwik-Way Co.)*

The expansion of the valve at operating temperature will cause seat contact to be closer to the center of the valve face. However, a valve seat positioned too high on the valve face will lead to burning near the margin.

The *lapping* of valve seats is done by placing abrasive compound between the valve face and seat and turning the valve back and forth (see Fig. 9-29). The heating of the valve and the resulting expansion is one reason for not *lapping* valve seats. Lapping provides a seat at room temperature, but as the valve expands, it moves away from the location where it was lapped. Occasionally lapping is required to obtain a suitable valve seat finish on seats that are difficult to machine, and it is even recommended by some manufacturers. It is recommended here that, if lapping is to be done, the seats be lapped with an old valve that has been refaced. In this way, the seat finish can be improved without undercutting

*FIGURE 9-29 Lapping a valve seat*

on the face of the valve to be installed in the engine.

The first step in seat grinding is to grind the primary seat angle, usually 45°, to clean base metal. Redress the grinding stone frequently to maintain good finishes. Grind the minimum required to remove all pitting. As with valve facing, pits must be removed, or burning will result. Remember, however, that grinding the 45° angle moves the valve deeper into the cylinder head, and this could cause problems when valve springs are installed or stem height above the spring seat is checked.

The second step is to grind a top angle of 30° (or the angle specified, if different). This smaller angle can be used to move the outer edge of the valve seat toward the center of the valve face.

The third step is to grind with a 60° stone (or the angle specified, if different). The larger angle can be used to narrow the seat to specifications and to make the seat even in width.

Three grinding angles are required if width, position, and uniformity are to be obtained (see Fig. 9-30). These requirements are most critical when longevity is desired. Somehow the notion has become popular that this seat grinding procedure is for "racing valve jobs." However, it is the passenger car, not the race car, that is expected to run up to 100,000 miles.

Seat position on the valve face can be checked by applying Prussian blue paste to the valve face and turning the valve slightly on the seat. When the valve is removed, the seat position can be read on the valve face (see Fig. 9-31). Be sure to turn the valve only slightly so that the relationship of the seat, valve, and valve guide can be checked. If contact does not appear all the way around the valve face, machining is unsatisfactory. If the valve is turned all the way around during this check, a contact pattern will be made all the way

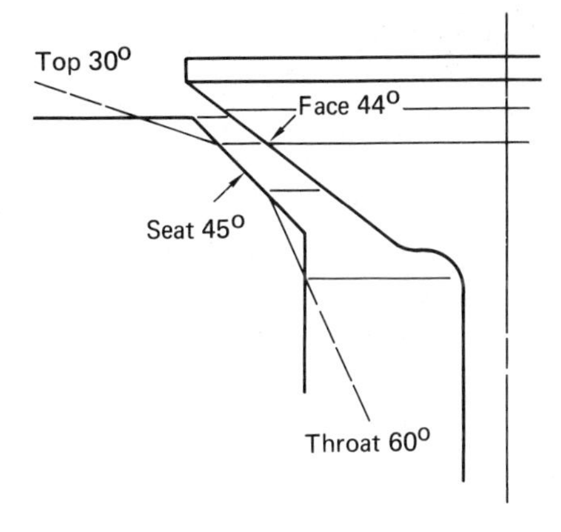


*FIGURE 9-30 A three-angle valve seat*

*FIGURE 9-31 Reading valve seat position on the valve face using Prussian blue paste*

*FIGURE 9-32 Using a dial indicator to check valve seat concentricity (Courtesy of Kwik-Way Co.)*

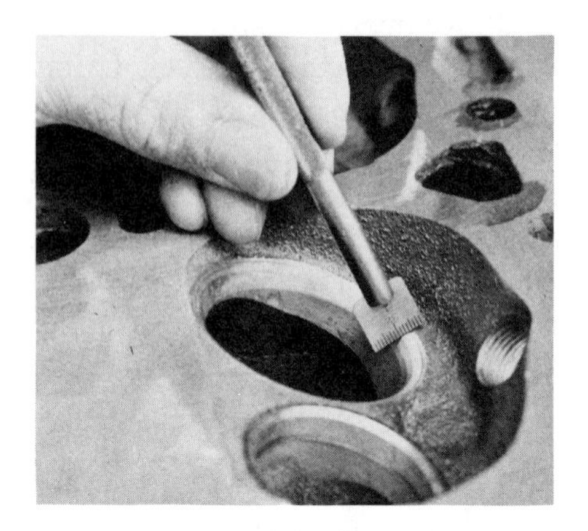

*FIGURE 9-33 Reading valve seat width with a scale*

*FIGURE 9-34 Sioux solid pilot (available in increments of 0.001 inch) and expandable pilot*

around the valve even if machining is incorrect. This check can also be made with a dial indicator (see Fig. 9-32).

The seat width is best read with a scale at the seat (see Fig. 9-33). The contact pattern made with the Prussian blue paste does not accurately show seat width because of the interference angle.

Stones are available in different diameters, angles, and compositions. Stones may be selected from catalogs according to materials. Pilots are available in common valve guide diameters and, if not adjustable, in increments of .001 inch (.025 mm) over size (see Fig. 9-34).

## INSTALLING VALVE SEAT INSERTS

Valve seat inserts are used to replace original equipment inserts that are no longer serviceable. They are also installed in cylinder heads having integral valve seats originally. The integral valve seats may be worn beyond service limits, or seat inserts may be installed as a part of a crack repair (cracks frequently extend across valve seats). Unless cracks are involved, the need to replace valve seats is generally discovered during seat grinding. At that time, it is found that to obtain a good seat finish, the seats must be ground until they are recessed too far into the cylinder head.

Frequently there are no specifications to use for determining at what point a seat should be replaced. It is recommended here that if valve spring installed height cannot be corrected with a .060 inch (1.5 mm) shim, a valve seat insert should be installed. Be sure to measure installed height of the valve springs with a new valve, because of the thicker margin, to minimize the measurement (see Fig. 9-35).

*FIGURE 9-35 Measuring the valve spring installed height with a telescoping gauge. This dimension is also referred to as assembled height.*

Although the installed height of springs may be corrected with shims of .060 inch (1.5 mm) and more, keep in mind that the valve stem height extending above the cylinder head will also become greater. As mentioned in the section on engine disassembly, excessive stem height will interfere with normal rocker arm and valve lifter operation and possibly hold valves open. It is also true that valve stems can be faced to reduce the stem length; however, they sometimes cannot be faced enough to restore the designed rocker arm geometry. In fact, some valve stems can be faced very little before the rocker arms interfere with valve spring retainers.

Valve seat inserts are available as replacements for original inserts or for repairing cylinder heads with integral seats. They will be listed in parts catalogs by engine application or by dimensions and materials. The dimensions used are outside diameter, inside diameter, and depth. The available materials include cast iron, hardened cast iron, and high nickel-

*FIGURE 9-36 Using an oversize seat cutter may lead to cutting through to the water jacket*

*FIGURE 9-37 An undersized valve welded to a seat as a driver*

*FIGURE 9-38 A bead welded around a seat insert to cause shrinkage*

chrome alloys. The cutter diameter required for installing each seat and the specified interference fit of the replacement seat is also given.

Use seats recommended for the particular engine whenever they are available. Other methods—such as selecting seats by size, usually to match cutter sizes on hand—frequently lead to cutting into water jackets (see Fig. 9-36). The recommended seats are also matched to the valve material. Valve and seat materials, even of the heaviest duty available, will not hold up in service unless the materials are properly matched.

The procedure for repairing a cylinder head with original seat inserts begins with removing the nonserviceable seats. If possible, remove the original seat without damaging the bore that it comes out of. For alloy and hardened seats, this may be done by welding the seat to an old valve of a smaller diameter. After being welded, the valve may be used as a driver for removing the seat (see Fig. 9-37). Another method for alloy and hardened seats is to weld a bead around the inside of the seat. The bead will shrink on cooling and loosen the seat so that it can be easily pryed out (see Fig. 9-38). Cast iron seats can be removed by using a cutter smaller than the replacement seat. The original seat is cut away to a thin shell and pryed out. The cutters will not cut hardened seats, so this method is usable only on soft seats.

Seat inserts can also be removed by prying them out with a special hooked pry bar (see Fig. 9-39). The problem with the pry bar method is that there is not always a place accessible for hooking onto the seat insert. It is also more difficult to remove the seat without damaging the bore.

Once the valve seat is removed, inspect the bore in the head for damage and then measure the inside diameter. Compare the in-

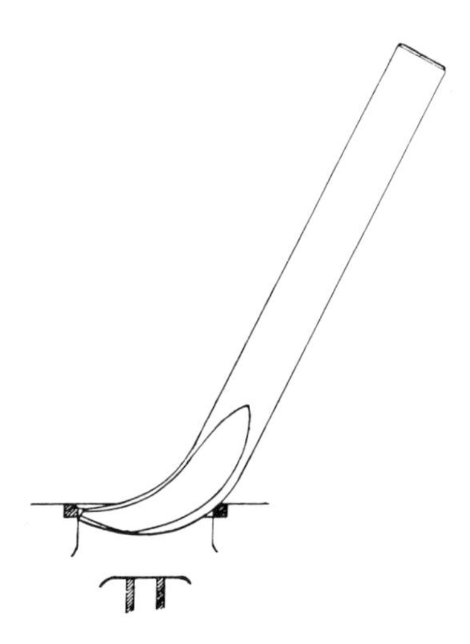

*FIGURE 9-39 A hooked pry bar used to remove a seat insert (Courtesy of K. O. Lee Co.)*

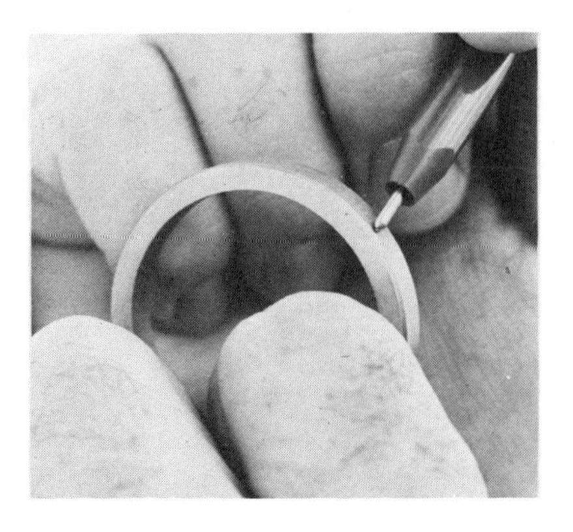

*FIGURE 9-40 Note the radius, or chamfer, on the lower corner of a seat insert*

side diameter of the bore to the outside diameter of the replacement seat. The difference in diameters should be within the range of specified interference that is shown in the specifications.

Use care when installing seat inserts. The bore should be clean, and an extreme pressure lubricant should be used to prevent galling. The radius, or chamfer, on the lower outside corner of the seat insert (see Fig. 9-40) must face down on installation to prevent broaching the walls of the bore. It is helpful on aluminum heads, because of greater interference fits, to heat the cylinder head to 200°F to 300°F. A driver and hammer are used to install the seat (see Fig. 9-41).

If it is found that the interference fit with a standard replacement seat is below specifications, an oversize seat must be used. Check the parts catalogs for the availability of oversize seats for a particular engine. If oversizes are not listed, it is sometimes recommended that the original bore be used because

*FIGURE 9-41 Installing a seat insert with Van Norman tools*

reboring may break into a water jacket. To reuse the original bore, look up another seat with the inside diameter, depth, and material the same as the original insert. The outside diameter should be oversize so that it can be cylindrically ground or turned down to just the right amount of interference.

The procedure for replacing integral seats with seat inserts is relatively simple compared to replacing original inserts. However, specialized automotive machine tools are generally used. The machining for seat inserts may be done on the same machine used for fitting false valve guides.

*FIGURE 9-42 The pilot used to guide a Van Norman seat cutter*

First, look up the seat insert called for and the required cutter. Second, select a cutter pilot which fits the valve guide (see Fig. 9-42). Install the pilot in the guide, place the cutter over the pilot, and engage the machine spindle to the cutter (see Fig. 9-43). Set the depth stop to the thickness of the seat insert (see Fig. 9-44). Set the spindle speed to the correct RPM. Go ahead with the boring operation.

*FIGURE 9-43 The pilot, cutter, and machine spindle used to cut an insert*

The finish bore is cleaned and lubricated so that the seat may be driven into place. The installed insert may be "staked" as an added precaution against the seat's coming out (see Fig. 9-45). In fact, the bore may be machined to an extra depth of approximately .005 inch (.13 mm) to make staking more effective. However, practice regarding staking operations varies considerably; and in many cases, seats are not staked at all, especially cast iron seats which have the same expansion rate as cast iron cylinder heads.

If both intake and exhaust seats are to be inserted, it is recommended that the intake be inserted first. The intake insert is usually of a softer material and the exhaust insert is

*FIGURE 9-44 Setting the depth of the cut to the thickness of a seat insert in a Van Norman machine*

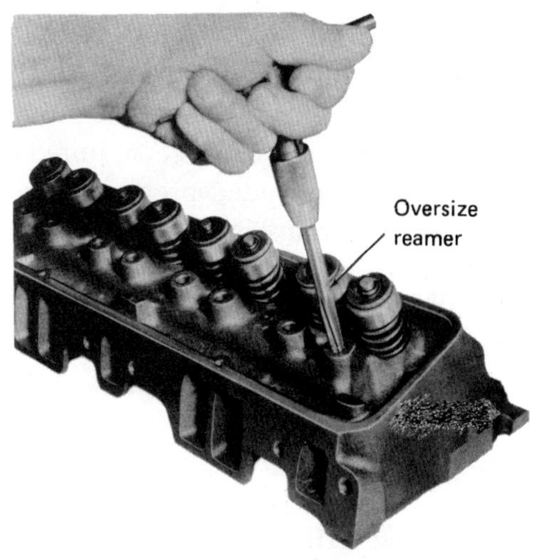


*FIGURE 9-55 Reaming the stud bore for the oversize replacement stud (Courtesy of Buick Motor Division)*

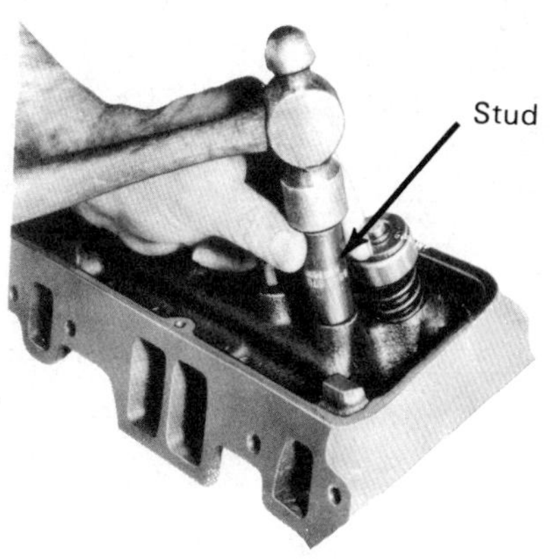


*FIGURE 9-56 A driver is used to install the stud to the correct depth. (Courtesy of Buick Motor Division)*

Threaded studs are also available (see Fig. 9-57). They require that the bore in the head be tapped to a thread matching the lower end of the stud. A commonly used thread size is 7/16-14. Care must be taken to align the tap with the bore centerline. Fixtures are sometimes available. Threaded studs are installed with a sealer on the threads because the bores go straight into the water jacket.

*FIGURE 9-57 A threaded rocker arm stud*

## CORRECTING INSTALLED HEIGHT

After refacing valves and grinding valve seats, the valve moves through the cylinder head to a greater depth. The length of the valve spring is also extended. It is essential that valve spring installed height be restored to original specifications so that specified spring tension is maintained.

*FIGURE 9-58 The location of valve spring shims*

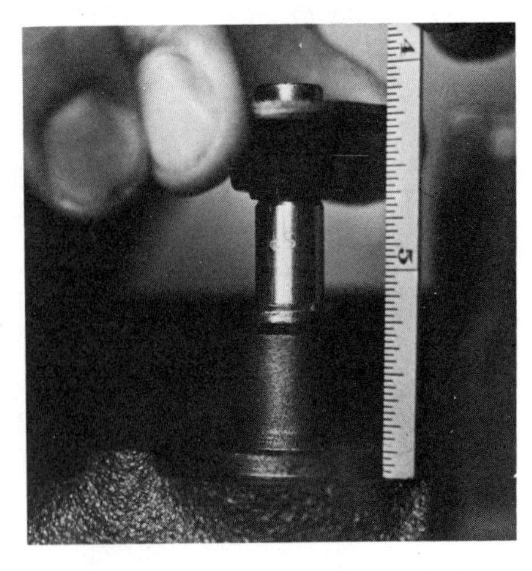

*FIGURE 9-59 Measuring installed height with a scale*

Installed height is corrected by placing shims under the valve spring (see Fig. 9-58). The installed height must first be measured to determine the thickness of shims required (see Fig. 9-59). A recommended procedure to follow for selecting shims is to use a .030-inch (.76 mm) shim, a .060-inch (1.52 mm) shim, or no shim at all, whichever brings the installed height nearest to specifications.

Keep in mind that aluminum cylinder heads already have a steel shim under the spring to prevent damage to the softer surface of the head. Keep this original shim in place when measuring installed height.

It is most important to note that shims are used to restore spring installed height to specifications—not to increase spring tension. Do not shim below installed height specifications. Spring failure can result because of added load on springs. Failure will surely occur if "coil binding" is created by compression of the valve spring to a point where coils bottom out against each other. Coil binding will bend push rods and flatten the camshaft.

## CORRECTING STEM LENGTH

Correct stem length, also called valve stem installed height, is measured from the spring seat to the top of the valve stem and is critical on engines with nonadjustable rocker arms. The length increases as a result of valve grinding procedures, and this lengthening can cause the hydraulic valve lifters to hold the valves open.

A recommended procedure is to measure the stem length on the valves at each end of the cylinder head (see Fig. 9-60). The length on these two valves can be corrected by facing the stems on the valve grinder. A straightedge may then be placed across the two end valves, and each of the remaining valves can be faced until they align with the straightedge. The stem length can be measured individually for each valve, but this tends to be somewhat slower.

Keep in mind that specifications for this dimension are not always readily available. Recall the emphasis on measuring stem length during disassembly of the cylinder head. Keep a notebook of these dimensions and you will save time.

If the head has been disassembled and the stem length is unknown, another procedure must be followed. First, the best that can be done is to make all stem lengths equal. The fastest way of doing this is by refacing the end valves the minimum required to obtain equal lengths. Then use the straightedge to check the lengths of the remaining valves and make corrections just as before. If valves remain open after following this procedure, it will be necessary to shim up rocker arm assemblies, use shorter push rods, or change to adjustable rocker arms. The exact solution depends on engine design and parts availability.

A real problem is created when no attention at all is given to stem length on engines with nonadjustable rocker arms. This oversight often leads to one or two valves that have long stems and, therefore, remain open. The open valves must be isolated and corrected. It may be necessary to remove the cylinder head again and correct the stem lengths. In the long run, it is faster to follow correct procedures and take the time to measure the valve stem lengths when they are disassembled.

*FIGURE 9-60 Measuring valve stem height above the head with a Hall-Toledo gauge*

## REFACING ROCKER ARMS

In many cases rocker arm wear on the surface that is in contact with the valve stem can be smoothed and refaced easily and quickly on a valve grinder. Some rocker arms formed or stamped from heavy-gauge steel cannot be refaced and are discarded if worn excessively.

Check the clearance between the rocker arms and rocker shaft before going ahead with the refacing operation. Sometimes the clearance is excessive, and both rocker arm bores and the shaft are worn. It is recommended that rocker arms be salvaged if clearance can be restored to specifications by replacing the shaft. Replace the entire assembly if replacing the shaft will not restore clearance to specifications. Excessive clearance between rocker arm and shaft will cause a reduction in oil pressure and should be corrected.

If rocker arms are to be salvaged, go ahead with the set up of the valve grinder and reface the rocker arms (see Fig. 9-61). During

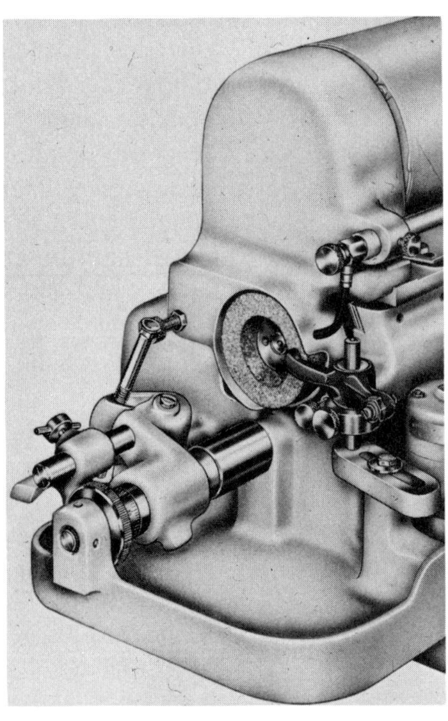

*FIGURE 9-61 Refacing rocker arms (Courtesy of Kwik-Way Co.)*

refacing, the grinding wheel is moved back and forth while the rocker arm face is held against it. A light pressure at the opposite end of the rocker arm keeps the rocker arm face in contact with the wheel. Grind only enough to clean up visible wear.

# 10

# Reconditioning Engine Block Components

Procedures for servicing engine parts will vary depending on the results of parts inspection and on the tools and equipment available. Recommended service procedures will also vary just as recommended service limits vary. The procedures discussed here are selected from among those practiced in the automotive engine service industry.

## CYLINDER HONING FOR OVERHAUL

After an engine has been run for any number of hours, a "glaze" forms on the cylinder walls. This glazed surface is hard and polished to a high quality surface finish. The glazed surface is unsuitable for the seating of new piston rings because the finish does not have sufficient abrasion to cause the rings to wear in.

An engine being overhauled typically has a certain amount of cylinder taper. Depending on the method and amount of cylinder honing, a small wear pocket will remain just under the ring ridge. Honing will change the quality of the surface finish for approximately 90 to 95 percent of the cylinder wall area. This percentage of cleanup is adequate for ring seating.

Many piston ring sets are sensitive to the quality of surface finish obtained from honing. For example, too fine a finish will not permit some chrome ring sets to seat properly. Too coarse a finish will cause some molybdenum ring sets to wear out at very low mileage. It is recommended that cast iron rings or ring sets specifically intended for engine overhauls be used when cylinders are not rebored. A properly honed cylinder will cause rings to seat within the first 500 miles. If ring seating does not occur within that mileage, it is possible that the cylinder finish and piston rings are poorly matched, and ring seating may not occur.

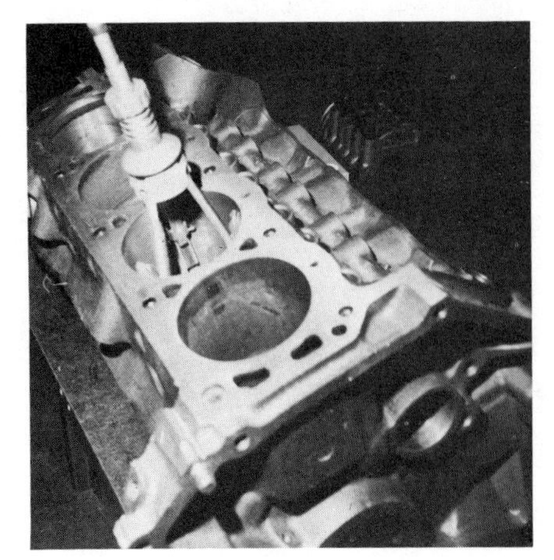

*FIGURE 10-1 Honing to break the cylinder glaze*

Portable tools are commonly used in general repair shops not specializing in automotive machining. One type of hone used is referred to as a *glaze-breaker* (see Fig. 10-1). These hones follow the shape and alignment of the *worn* cylinder, and care should be taken to hone the minimum required for proper glaze breaking.

## REBORING AND HONING CYLINDERS

It was recommended earlier that cylinders exceeding the limits for taper be rebored for oversize pistons. Common oversizes for domestic engines are .030 inch (.75 mm), .040 inch (1.02 mm), and .060 inch (1.52 mm) over standard size. Some .020-inch (.5 mm) oversize pistons are available on special order. Common oversizes for import engines are .50 mm, .75 mm, and 1 mm. The minimum oversize that will remove all cylinder wear should be used. This leaves the maximum cylinder wall thickness and permits another reboring later.

An oversize cylinder is typically finished to the exact oversize. That is, a 4-inch (101.6

*FIGURE 10-2 A Rottler boring machine. One or more machines of this type may be used to increase production*

mm) diameter cylinder being fitted with .030-inch (.76 mm) oversize pistons is machined to 4.030 inch (102.36 mm). This practice generally follows the recommendations of piston manufacturers. The measured piston diameter will be undersize from the finished bore diameter by the amount of specified piston clearance. Measure each set of pistons to be sure that this works out. An acceptable set of pistons will fit in the cylinders with minimum specified clearance up to an additional .001 inch (.025 mm).

Boring machines are used in production machine shops (see Fig. 10-2). Two boring bars may be used at a time. One is set up while the other is running. Maximum cuts are possible because of the rigidity of the setup. They are also faster to set up because of air-float and air-clamping devices. One distinct advantage is that the boring bars are positioned from a table above the block. The table is parallel to the main bearing bores, thus ensuring that cylinders are bored 90° to the crankshaft.

Portable boring bars are popular in smaller machine shops (see Fig. 10-3). These machines do a comparable job if used properly. Note that they locate from the block surface. This means that the surface must be cleaned and deburred before setting up the boring bar. These bars will not bore the cylinders 90° to the crankshaft unless the block surface is parallel to the main bearing bores. This is usually no problem, but sometimes block surfaces are out of alignment. One method of checking this is to measure the distance from the main bearing bores to the top surface of the block at each end of the block (see Fig. 10-4).

Methods of cylinder boring will vary according to final honing procedures. For example, if a cylinder is to be honed to final size by hand, the cylinder is bored to within .003 inch (.08 mm) of finished diameter. The

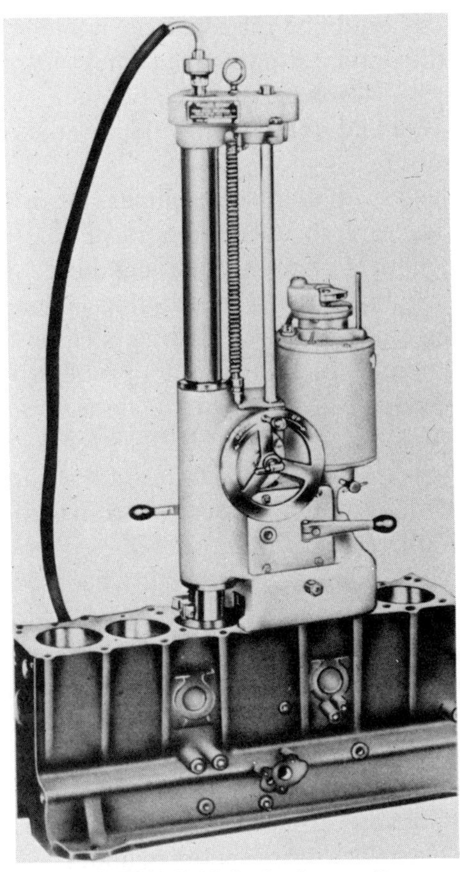

*FIGURE 10-3 Boring cylinders with a portable Kwik-Way bar (Courtesy of Kwik-Way Co.)*

*FIGURE 10-4 Measuring block parallelism*

recommended method is to rough bore first to .005 inch (.13 mm) under finished size. The cylinder finish is then improved by making a .002 inch (.05 mm) cut with a boring tool lapped to finish boring specifications.

Cylinder honing by hand will produce a high quality cylinder finish if done properly. The hone is a rigid type of hone that can be made to produce round, straight bores (see Fig. 10-5). Stones are selected according to the surface finish quality desired. It is sometimes recommended that rough honing be done dry and that honing oil be used for finish

*FIGURE 10-5 Honing by hand after boring cylinders*

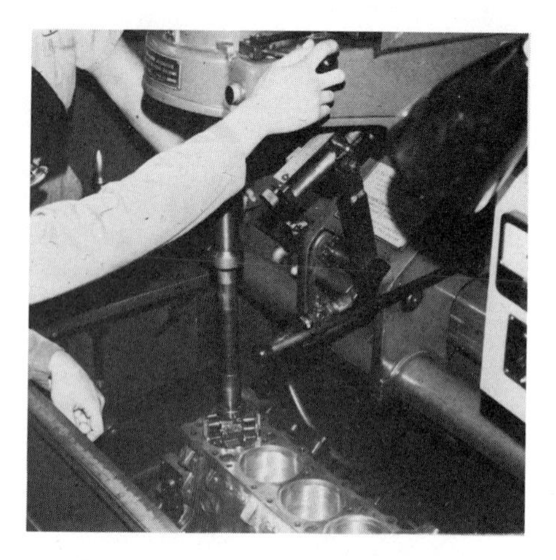

*FIGURE 10-6 A Sunnen CK-10 cylinder honing machine in operation*

*FIGURE 10-7 The chamfer at the top end of a cylinder*

*FIGURE 10-8 Chamfering with a grinder*

honing. The honing oil cools the cylinder and cleans the honing stones so that they cut more efficiently. However, the specific instructions of the tool and abrasive manufacturer should be followed.

The use of honing machines has certain advantages over hand honing (see Fig. 10-6). First, cylinders need only be rough bored .005 inch (.13 mm) undersize, if boring is done at all. Because honing by machine is easier, the extra amount of honing is not a problem. Cylinder finishes are obtained by abrasive selection, honing speed, and the stroking rate. Honing speed and stroking rates are a part of the machine setup and vary according to cylinder diameter and length.

The honing machines may be used to finish cylinders oversize without boring. A roughing abrasive is used to bring cylinders to within .003 inch (.08 mm) of finished size. Alignment of the cylinders remains the same as original. This is ensured by "dwelling," or beginning the rough honing in the lower (unworn) portion of each cylinder. The hone remains at the bottom of the cylinder, comes up the full length of the cylinder, and then "dwells" again. As soon as the cylinder wear is removed, honing is done continuously along the full length of the cylinder. Finishing stones are used to remove the last .003 inch (.08 mm). Eight cylinders can be honed .030 inch (.76 mm) oversize in less than an hour.

The top edge of each cylinder should be chamfered to a width of approximately 1/32 inch (.8 mm) (see Fig. 10-7). This is done so that piston rings will slip into the cylinders easily during assembly. The chamfering operation may be done by hand feeding a 60° tool into the edge of the cylinder as a last step after boring. It may also be done by hand grinding (see Fig. 10-8) because the exact angle and width of the chamfer are not critical.

A special emphasis must be made on obtaining correct surface finishes. Surface fin-

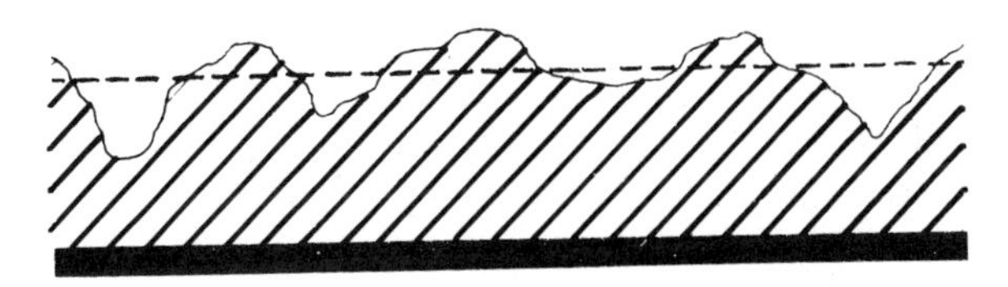


*FIGURE 10-9 An illustration of the principle of the RMS microinch surface finish scale (Courtesy of Sunnen Products Co.)*

ishes can be measured and placed on a comparative scale. One scale is root mean square (RMS) microinches. The numerical rating on this scale is equivalent to one third of the depth of surface variations (see Fig. 10-9). Chrome and cast-iron piston rings require an average surface finish of 25 RMS microinches. Moly piston rings require a surface finish of approximately 15 RMS microinches. A finish that is too fine will not have sufficient abrasion to seat chrome rings. A finish that is too coarse will cause moly rings to wear out very quickly. Be careful to follow the honing procedures recommended for the type of piston ring.

The included angle of the crosshatch pattern left on cylinder walls after honing is approximately 60° (see Fig. 10-10) and is sometimes thought to be an important aspect of surface finish. It has been found that the angle can vary somewhat without affecting piston ring performance. The angle is controlled in honing machines but must be observed visually during hand honing. If the angle is too small, increase the stroking rate. If the angle is too large, decrease the stroking rate.

*FIGURE 10-10 The crosshatch pattern in properly honed cylinders*

## CYLINDER SLEEVING

Piston ring breakage and piston seizure caused by engine overheating or lack of lubrication are common causes of cylinder scoring. Cylinders may also be cracked or broken by piston failure or by compressing coolant which has leaked into a cylinder.

Scored cylinders may be repaired by a number of methods depending on the extent of the damage, the overall engine condition, and the budget allowed for the repair. For example, a scored cylinder, if not scored too deeply, may be repaired by boring to an oversize. It is not unusual to find only one cylinder bored oversize. Although this practice may seem strange, engine operation will be perfectly normal providing the weights of the original and replacement pistons are closely matched. This would probably be the most economical repair.

In the case of a scored cylinder in a badly worn engine, it may be desireable to bore all cylinders oversize. If cylinder scoring should be too severe on one or two cylinders, those cylinders can be sleeved. However, the cost of sleeving, reboring, and pistons must be considered.

In cases of cylinder damage on low mileage engines, it is sometimes desireable to sleeve one damaged cylinder to match the diameters of the other serviceable cylinders. In this way, the cost of piston replacement, sleeving, and reboring is limited to perhaps only one cylinder.

Deeply scored or cracked cylinders can be repaired only by sleeving. A cylinder repaired by sleeving will have strength comparable to the original engine block. Sleeves are generally available in 3/32-inch (2.38 mm) or 1/8-inch (3.17 mm) wall thicknesses and in various lengths. Some metric sizes are also available. The sleeves come approximately .015 inch (.38 mm) undersize on the inside diameter so that the specified wall thickness is obtained only after finish boring and honing.

Boring for the sleeve requires going approximately 3/16 inch (4.76 mm) oversize for 3/32-inch (2.38 mm) wall sleeves and approximately 1/4 inch (6.35 mm) oversize for 1/8-inch (3.17 mm) wall sleeves. The bore diameter should allow for a .002-inch to .003-inch (.05 to .07 mm) press-fit of the sleeve.

A shoulder is sometimes left at the bottom of the cylinder to retain the sleeve. The shoulder is approximately 3/16 inch (4 mm) wide (see Fig. 10-11). The shoulder is formed when the power feed on the boring bar is disengaged short of passing through the cylinder. The top edge of the cylinder and the bottom edge of the sleeve should be lightly deburred after the final cut in preparation for installing the sleeve (see Fig. 10-12).

Methods of installing the sleeve vary. One method calls for using a sleeve approximately 1/8 inch (3 mm) longer than the cylinder. The sleeve is then driven into place using a hammer and driver (see Fig. 10-13). The sleeve then extends approximately 1/8 inch (3 mm) above the top surface of the block after installation. The extra length is removed by hand feeding a flat cutting tool in a boring bar down until the surface of the sleeve nearly blends with surface of the block. Final cleanup is then made by draw-filing. Extra sleeve length may also be cut off with a saw before or after installation, and the block surface trued by resurfacing after the sleeve is installed.

Another method of installation is to chill the sleeve in a bucket of dry ice and kerosene to shrink the outside diameter. The combination of dry ice and kerosene chills the sleeve to approximately −30°F without the formation of ice. The kerosene also serves as a lubricant. The block is heated to 200°F to expand the bore diameter. A temperature difference of approximately 175°F between the sleeve

*FIGURE 10-11 A block cutaway showing the shoulder at the bottom of a cylinder used to stop a cylinder sleeve*

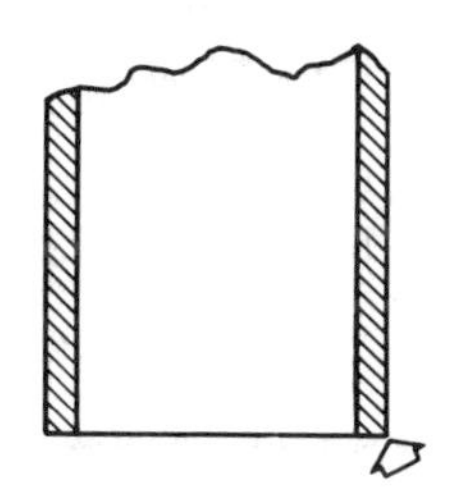

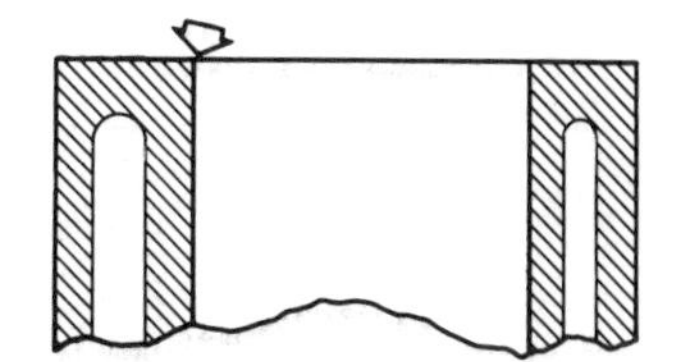

*FIGURE 10-12 Where to deburr cylinders and sleeves prior to installation*

*FIGURE 10-13 Driving in a cylinder sleeve*

*FIGURE 10-14 A Tobin Arp line boring machine set up for main bearing housing bores (Courtesy of Tobin Arp Mfg. Co.)*

and the block will change the press-fit to a slip-fit in all cases except perhaps the largest diameter sleeves. The sleeve is held securely in place once temperatures equalize. Even if a slip-fit is not obtained, the amount of press-fit is reduced and installation is made easier.

## ALIGN BORING OR HONING

There are several causes for misalignment of main bearing bores. The blocks warp, the caps stretch, and sometimes the caps are replaced. The alignment of the bore centerline and the housing bore diameters are corrected by align boring (see Fig. 10-14) or by line honing (see Fig. 10-15).

The first step in preparing an engine block for boring or honing is the grinding of the main caps. These caps are ground at the parting line to reduce the housing bore diameters

*FIGURE 10-15 A Sunnen line honing machine in operation*

.002 inch (.05 mm) or more below specifications (see Fig. 10-16). This permits remachining to original bore specifications. Particular attention should be given to grinding the cap for the thrust bearing. Grinding must be done on exactly the same plane as the original parting line so that the cap does not tilt in assembly. Any tilting of this cap will decrease crankshaft end play.

In line honing, the honing is done through all main bearing bores at the same time. This means that the centerline position will be an average of all bore positions. The centerline also moves down into the block very slightly. The amount of change is approximately half the reduction below specified bore diameters after cap grinding. *This is not equal to the amount ground from the caps:* the caps may have been stretched. Some care should be taken to reduce bore diameters to exactly .002 inch (.05 mm) undersize, meaning that the centerline changes .001 inch (.025 mm). This amount of change in centerline position causes no problems.

Align boring, as the name implies, involves setting up a boring bar and a single pointed tool through the main bearing bores. The boring bar is usually positioned from the two end housing bores. This means that alignment of the centerline is determined by these two bores. As with line honing, the centerline moves down into the block very slightly. However, with align boring, the amount ground from the caps has no influence on the change in centerline position. This is because the boring bar can be positioned from the block side of the bores, and the caps have no influence on the boring bar position. Setups are made very carefully so that the depth of cut on the block side of the bore is minimal.

There are advantages to each machining process. Line honing is fast and precise. It is

*FIGURE 10-16 Grinding the parting face of main caps on a Sunnen cap and rod grinder*

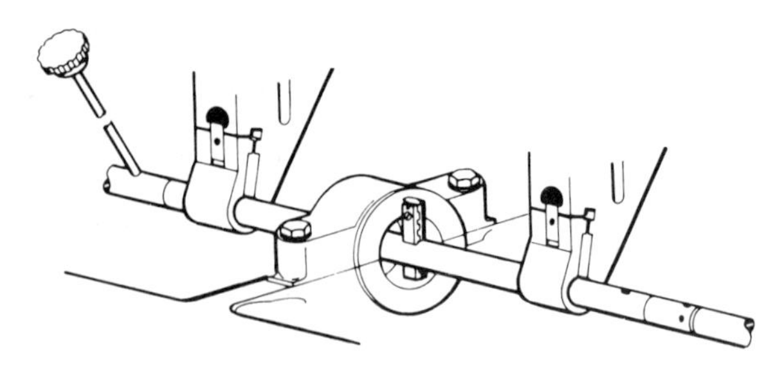

*FIGURE 10-17 Facing main bearing thrust faces in a Tobin Arp line boring machine (Courtesy of Tobin Arp Mfg. Co.)*

ideally suited to correcting problems of warpage and stretching of the caps. Align boring is better able to handle severe misalignments such as when caps are changed. Multiple-diameter main bearing bores can only be remachined by boring. In addition, the facing of thrust surfaces on a bearing can only be done in align boring machines (see Fig. 10-17).

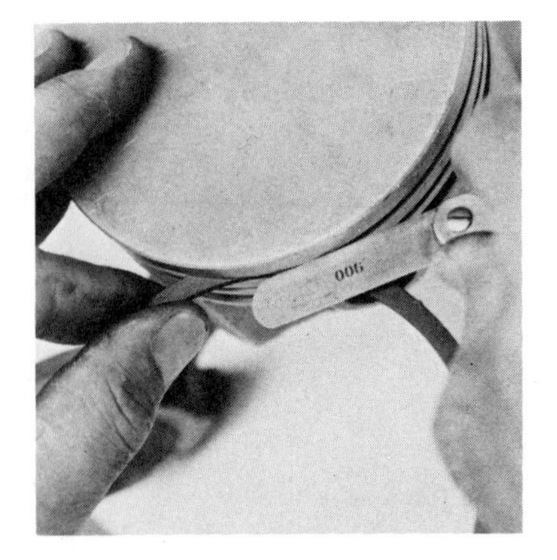

*FIGURE 10-18 Checking for ring groove wear with a ring and feeler gauge*

## REGROOVING PISTON RING GROOVES

Wear of the top compression ring groove is a common occurrence. As mentioned before, the groove may be checked for wear with a gauge for the purpose (available from ring manufacturers) or with a new piston ring and a .006-inch (.15 mm) feeler gauge. If the gauge or the piston ring and .006-inch feeler gauge can be inserted into the groove, the condition is unacceptable (see Fig. 10-18).

The ring groove is recut using a tool of the standard ring groove width plus the thickness of a spacer (see Fig. 10-19). The new ring is then fitted to the oversize ring groove by using a steel spacer on top of the piston ring to take up the clearance (see Fig. 10-20). Ring groove spacers are available in .024-inch (.61 mm) and .031-inch (.79 mm) thicknesses. Failure to correct ring groove wear can result in increased oil consumption. The oil consumption is caused by "oil pumping." That is, oil accumulated under compression rings on the down stroke is squeezed around the top of the rings on the upstroke (Fig. 10-21).

*FIGURE 10-19 Recutting a ring groove using a Perfect Circle Manulathe*

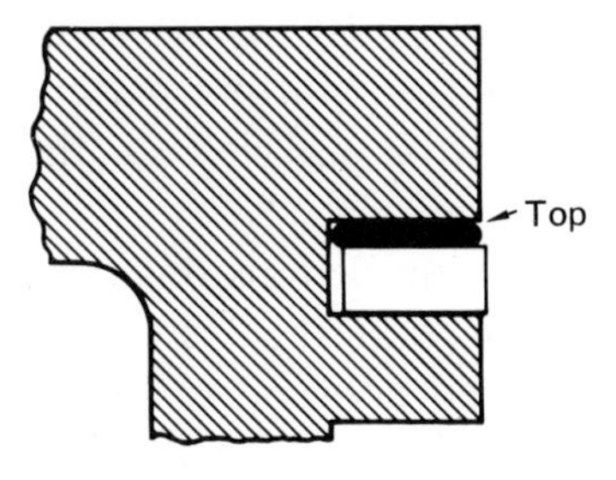


*FIGURE 10-20 A ring groove spacer installed after the ring groove is recut*

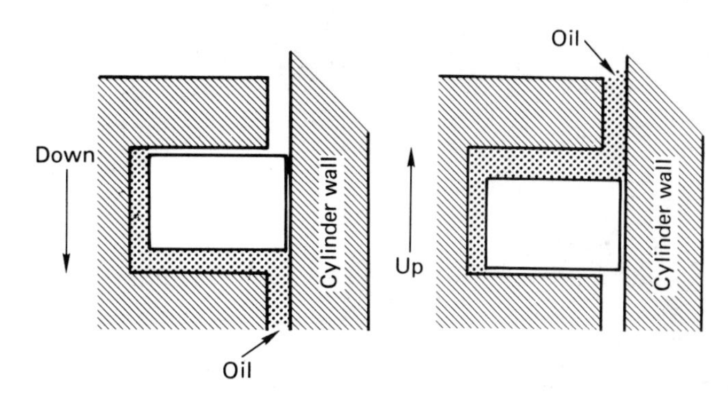


*FIGURE 10-21 "Oil Pumping" past rings due to increased side clearance of piston rings*

It is standard practice to limit recutting ring grooves to the top ring groove. When lower ring grooves are worn, it is likely that the engine needs reboring and the pistons need replacing. Recutting the lower ring grooves would also weaken the piston ring lands.

## PISTON KNURLING

It is often found that after worn cylinders are honed for overhaul, piston-to-cylinder wall clearance is excessive. Piston noise will result as well as piston ring wear because of the rocking of the piston in the cylinder (see Fig. 10-22).

Piston-to-cylinder wall clearances up to approximately .006 inch (.15 mm) can be corrected by expanding the piston diameter by knurling (see Fig. 10-23). The machines used

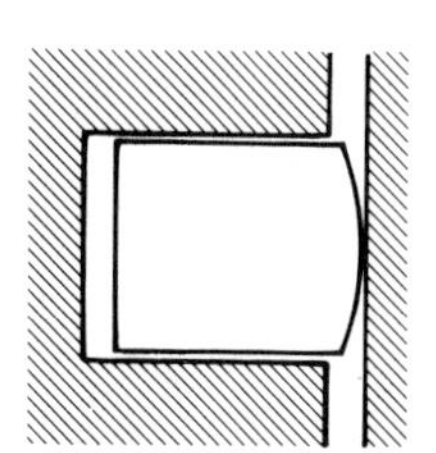

*FIGURE 10-22 "Barrel-faced" rings caused by rocking of the piston and ring in the cylinder.*

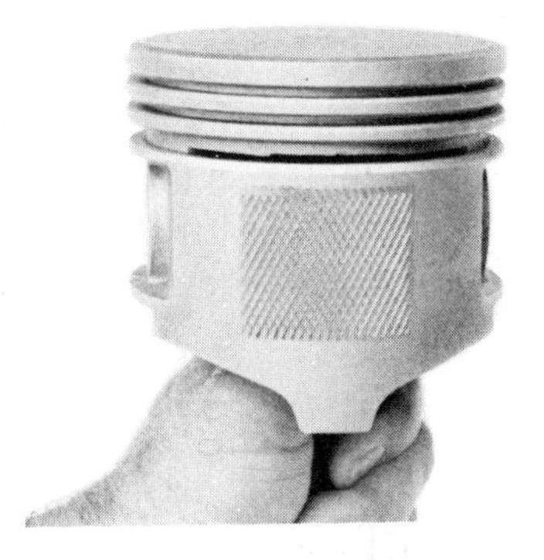

*FIGURE 10-23 A piston expanded by knurling*

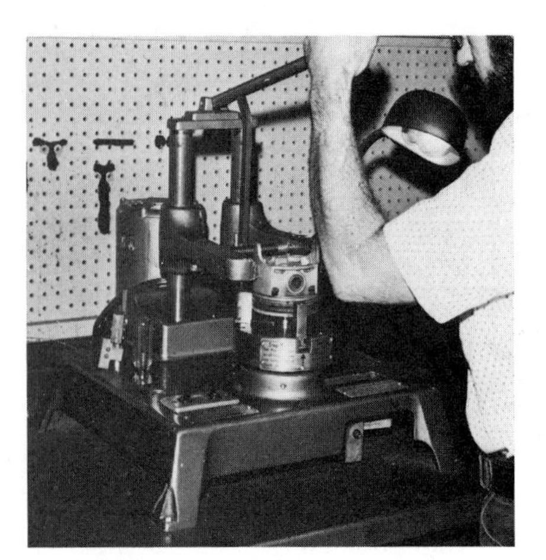

*FIGURE 10-24 Knurling a piston in a K-Line piston knurler*

for piston knurling (see Fig. 10-24) are capable of varying how far the knurling pattern extends around the piston as well as the amount of piston expansion.

Practices for fitting knurled pistons vary. One recommendation is that the pistons be knurled oversize and then polished to fit the cylinders. Knurled pistons are fitted with tight clearances because of the reduced surface area in contact with the cylinder wall. Clearances are usually no more than half the specified clearance.

## FITTING OVERSIZE PISTON PINS

Oscillating piston pins are the most commonly used type in newer engines. With this design, the piston pin is press-fit through the connecting rod with approximately a .001-inch (.02 mm to .03 mm) interference. The piston pin fits with approximately a .0005-inch (.01 mm) clearance through the pin bores in the piston.

As mentioned earlier, it is difficult to determine the piston pin clearance in assembly. Therefore, direct measurements are used only when the pistons are removed from the rods. Removing the pistons from the rods may be included in an overhaul because piston pin noises were diagnosed. It may also be decided to disassemble pistons and rods during the process of overhaul when obviously loose pin fits are found by "feel." This is a particular problem because loose-fitting piston pins are sometimes quiet in a worn engine because pistons slide freely in the cylinder but become noisy when the increased friction of new rings cause the pistons to slide less freely.

*FIGURE 10-25 Honing a rod bore for an oversize piston pin*

Using new standard pins or installing oversize pins are the accepted methods of eliminating pin noises. The connecting rod pin bores must be honed oversize to obtain the approximate .001-inch (.02 mm to .03 mm) press-fit for the oversize piston pins (see Fig. 10-25). The piston pin bores are honed to the

*FIGURE 10-26 Honing piston bores to the correct clearance for oversize pins*

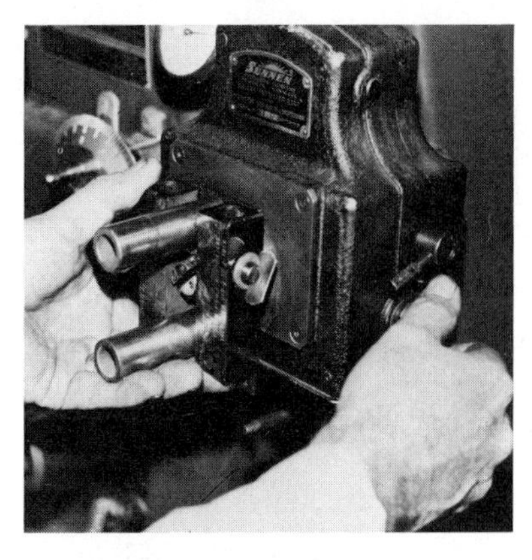

*FIGURE 10-27 Setting a Sunnen AG300 gauge to the piston pin diameter*

specified clearance using a honing unit with sufficient length to hone the two bores simultaneously (see Fig. 10-26). This method aligns the bore centerlines.

Highly accurate measuring methods are required to check clearances as close as those specified. Sunnen manufactures a precision gauge that is set to the piston pin diameter (see Fig. 10-27). The gauge can then be used to measure the amount of interference in the connecting rod (see Fig. 10-28) and the amount of clearance in the piston pin bores (see Fig. 10-29). Checking new pin fits by feel, although

*FIGURE 10-28 Gauging the rod bore to check for interference*

*FIGURE 10-29 Gauging piston pin bores to check clearance*

fairly common, is highly unreliable for the small clearances specified.

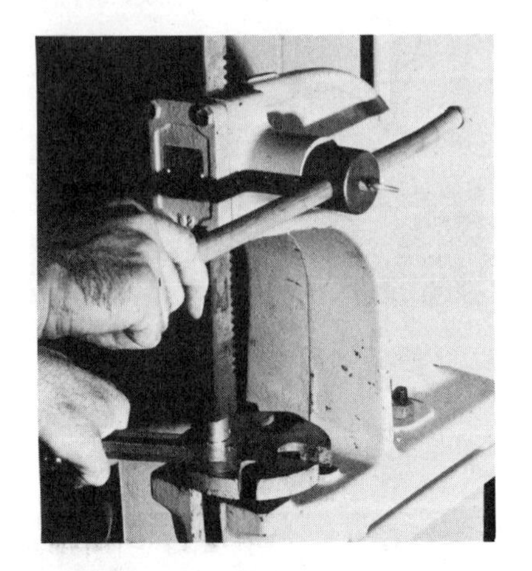

*FIGURE 10-30 Pressing a bushing into a rod for a full-floating piston pin*

## FITTING FULL-FLOATING PISTON PINS

Full-floating piston pins permit the free rotation of the piston pin through bores in the piston and the rod. Wear of the piston pin is minimized because the wear is distributed around the circumference. In fact, the fitting of full-floating piston pins is often limited to replacing and resizing the bushings in the connecting rods. This is done as required to reduce pin noise in overhauled engines and as a part of the more complete procedures used in engine rebuilding.

The first step is to press out the old bushings. Replacement bushings are then pressed into place (see Fig. 10-30). Predrilled oil holes should be pressed in place so that they align with the holes in the connecting rod. If oil holes are not predrilled, they should be drilled after the bushing installation and before finish machining.

Bushings may be expanded and seated in place using an expanding tool in a precision hone (see Fig. 10-31). This procedure is recommended because the bores in the connect-

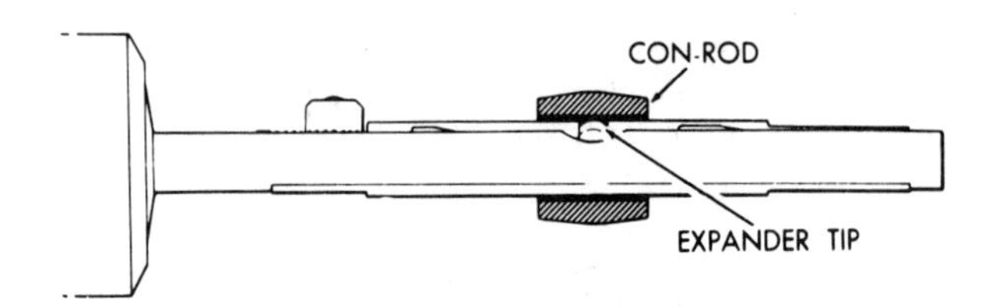


*FIGURE 10-31 Expanding a rod bushing into place with the Sunnen bushing expander (Courtesy of Sunnen Products Co.)*

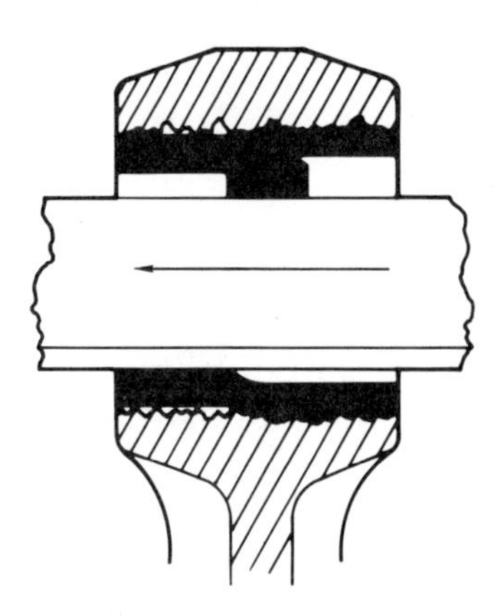

*FIGURE 10-32 Expanding a rod bushing seats the bushing into the rough-finished bore (Courtesy of Sunnen Products Co.)*

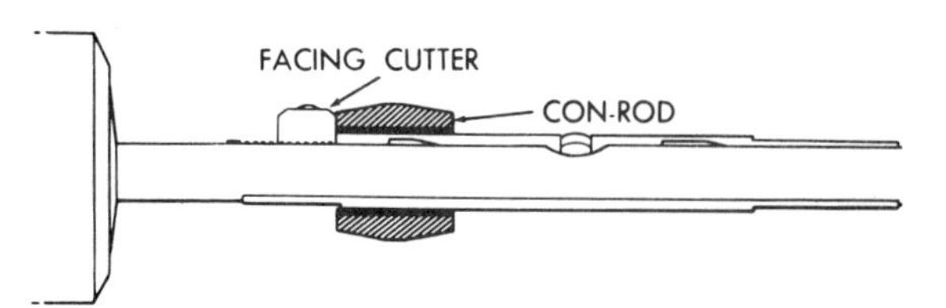


*FIGURE 10-33 The Sunnen bushing expander is also equipped with a facing tool for the rod bushings (Courtesy of Sunnen Products Co.)*

ing rods are often rough and not suited to holding the bushing in place (see Fig. 10-32). This same tool may be used to face off any bushing material extending from the bore (see Fig. 10-33). Bushings with steel backs should not be expanded.

The bushings are first rough honed undersize with a coarse abrasive stone. The stone is then changed, and the bushing is honed to size. Keep in mind that precision gauging is recommended to obtain specified clearances of approximately .0004 inch (.01 mm).

Piston pin bushings are also resized in a boring machine (see Fig. 10-34). These boring machines have advantages. One is that the centerline of the bushing is bored parallel to the centerline of the rod housing bore. This alignment corrects for bend or twist in the connecting rod and eliminates the need to align rods. Boring machines also have provisions to maintain center-to-center spacing of rod bearing and piston pin bores.

*FIGURE 10-34 Boring a rod bushing in a Tobin Arp machine*

## ASSEMBLING PISTONS TO CONNECTING RODS

The first step in preparing for assembly is to determine the position of each connecting rod and piston. Most pistons have a notch, an arrow, the letter *F,* or the word FRONT on the piston. These marks indicate that the piston is to be installed with the mark pointing to the front of the engine. Connecting rods may be positioned in respect to an oil spurt hole, a chamfer on the rod housing bore, rod numbers, or bearing tangs (see Fig. 10-35). A typical assembly, for example, calls for the notch in the piston to face the front and the connecting rod oil spurt hole to point toward the camshaft.

Keep in mind that all engines are not alike. Correct assembly for one engine may be

*FIGURE 10-35 Typical instructions for the correct installation of connecting rods and pistons (Courtesy of Chrysler Corp.)*

incorrect for another. Note also that assemblies for most V-block engines will be correct for only one side. Check the order of assembly for these engines carefully because pistons and connecting rods are assembled differently for left and right banks.

Incorrect assembly may cause the piston pin offset in the piston to be on the wrong side of the cylinder and result in piston failure. Keep in mind that the piston pin is offset to the side of the cylinder opposite the crankpin on the power stroke. This keeps the piston from rocking severely at the top of the stroke, thereby reducing engine noise and extending piston life. Valve reliefs may also be out of position in respect to the valves, causing piston and valve interference. Connecting rods out of position may cause insufficient cylinder lubrication because of the direction of oil spurt holes. Connecting rod bearings are also usually offset in V-block engines. The offset is toward the center of the crank pin so that bearings will not ride on the fillets at either edge of the crank pin.

The assembly of full-floating pistons and connecting rods is quite easy once the correct order is determined. Because full-floating designs use a slip-fit of the piston pin through the piston pin bores and connecting rod bushing, all that is required is cleaning and lubrication before slipping the parts together. It is recommended that locking rings be replaced as a precaution against failure each time the engine is disassembled.

The assembly of oscillating piston pins is somewhat more painstaking. As with full-floating designs, parts should be clean, lubricated, and laid out in order of assembly. The problem is that the piston pin is a press-fit. The necessary tools may be available from the engine manufacturer or tool manufacturers serving the repair industry. Cautions include

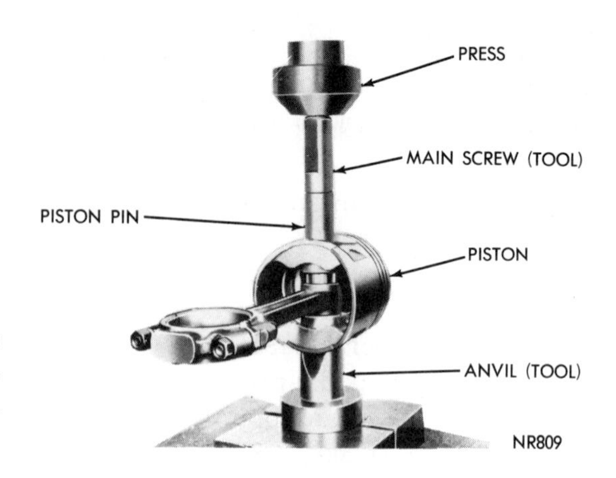


*FIGURE 10-36 Assembling a piston to a rod using the manufacturer's tools (Courtesy of Chrysler Corp.)*

lubricating the piston pin to prevent galling as it moves through the rod and to use the correct tools and adapters so that damage or distortion of the pin bores in the piston is avoided (see Fig. 10-36). If pin bores are chamfered on one side only, take care to press pins into the chamfered side first.

The piston pin must also be pressed through from one side only so that the connecting rod is centered on the piston pin. Efforts to correct the position of a piston pin that has been pressed through too far by pressing from the opposite side frequently results in distortion of the piston.

An alternate method to pressing oscillating piston pins into assembly is to use heat to expand the rod. Equipment is available to preheat the "eyes" of the rods to approximately 425°F to change the press-fit of the piston pins to a slip-fit (see Fig. 10-37). The piston pin at room temperature is quickly inserted by hand through the heated rod.

*FIGURE 10-37 Using a Sunnen CRH 50 rod heater to expand the eye of the rods for assembly of the pistons and rods*

## RESIZING CONNECTING ROD HOUSING BORES

As mentioned before, rod housing bores frequently become stretched .001 inch (.02 mm or more) or more oversize. It is necessary at this point to restore the housing bore to original specifications for roundness and diameter to be assured of normal bearing service.

The reconditioning procedure is begun by removing the connecting rod bolts. This is done by clamping the rod in a vise and driving out the bolts with a brass hammer (see Fig. 10-38). In some situations, the bolts and nuts are discarded at this point so that they can be replaced with new parts when the rods are reassembled. If magnafluxing is to be done, it should be done while the bolts are removed.

The next step is to grind approximately .002 inch (.05 mm) from the parting faces of both the rod and the cap (see Fig. 10-39). It will occasionally be found that the parting line between the rod and cap is badly out of alignment. In such cases, more material may have to be ground from the parting faces to obtain a clean, flat surface.

The housing bore can be resized in power-stroking honing machines (see Fig. 10-40). Note that the rod is placed over the honing unit and inside a power-stroking attachment.

*FIGURE 10-38 Driving out rod bolts using a brass hammer and a rod vise*

*FIGURE 10-39 Grinding the parting faces of rods and caps in a Sunnen cap and rod grinder*

*FIGURE 10-40 Resizing connecting rod housing bores in a Sunnen power-stroking hone*

*FIGURE 10-41 Hand honing connecting rod housing bores in Sunnen hone without the power-stroking attachment*

*FIGURE 10-42 Boring the connecting rod housing to size in a Tobin Arp machine*

The speed of rotation of the hone and the stroking rate are adjusted according to the housing bore diameter. The correct combination of honing speed and stroking rate produces a round, straight bore.

Rods are honed two at a time when honed without a power-stroking type of machine (see Fig. 10-41). Placing two rods side by side and periodically reversing their position helps the operator obtain a round, straight bore. In hand honing, only honing speed is adjusted according to housing bore diameters.

Rods may also be bored to size (see Fig. 10-42). The rod housing bore is bored on a centerline parallel to that of the piston pin. This eliminates the need for aligning the rod later.

The center-to-center distance between the housing bore and the pin bore is shortened after the housing bore is resized. This situation is comparable to align boring or line honing. The amount of change is approximately half the amount of undersize in the housing bore after cap and rod grinding and before resizing. While the amount of change is slight, the bor-machine has advantages in this respect. The center distances may be bored at least to match. Rods with piston pin bushings may also be bored to specified center distances by replacing the bushings and boring through the bushing as shown earlier. This moves the bushing centerline slightly but also corrects the center distance.

New housing bore specifications call for roundness within .0003 inch (.008 mm). Housing bores are resized to the midrange of the specified diameter and within the limits for roundness. Because these specifications are very close, precision gauging of some type is required (see Fig. 10-43).

Other minor machining operations may be done during connecting rod reconditioning. One of these is the deburring of the sides of

*FIGURE 10-43 Gauging the rod housing bore in a Sunnen AG300 gauge*

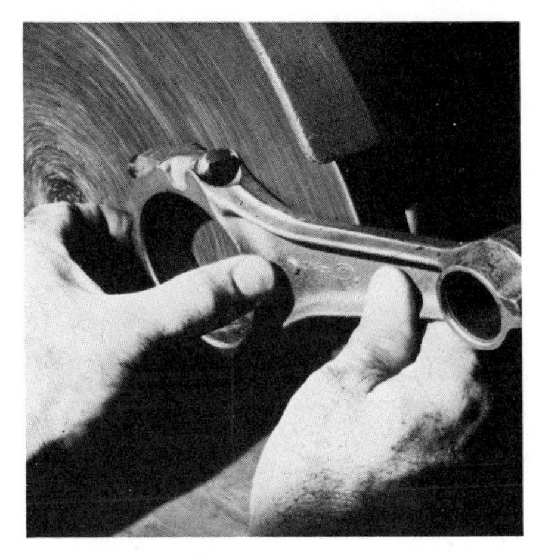

*FIGURE 10-44 Deburring the sides of the connecting rods*

the rods by lightly sanding the surfaces (see Fig. 10-44). Be careful not to narrow the rods excessively, or oil throw-off will be increased. Another operation is the rechamfering of the rod housing bore. This is done by replacing the honing stone with a chamfering tool in the same setup used for honing the housing bores (see Fig. 10-45). The deburring operation prevents scuffing between rods on the crankshaft. Rechamfering restores oil throw-off characteristics as designed by the manufacturers.

*FIGURE 10-45 Chamfering rod housing bores with a Sunnen chamfering tool*

## ALIGNING CONNECTING RODS

Rod alignment is generally checked after the required machining operations are completed. Two conditions are frequently found. One condition is called *bend* and the other is called *twist* (see Fig. 10-46). Both conditions cause

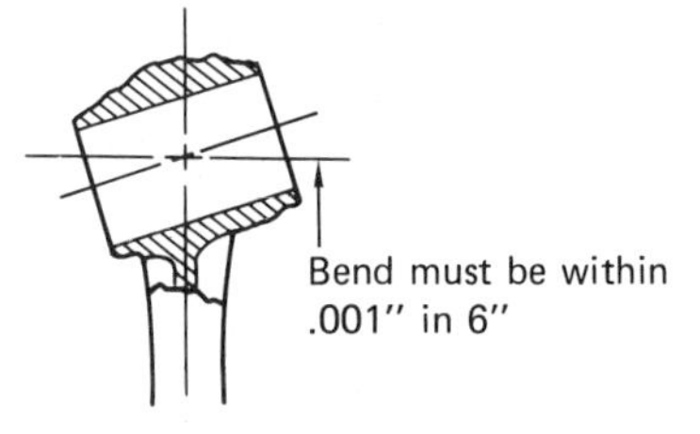


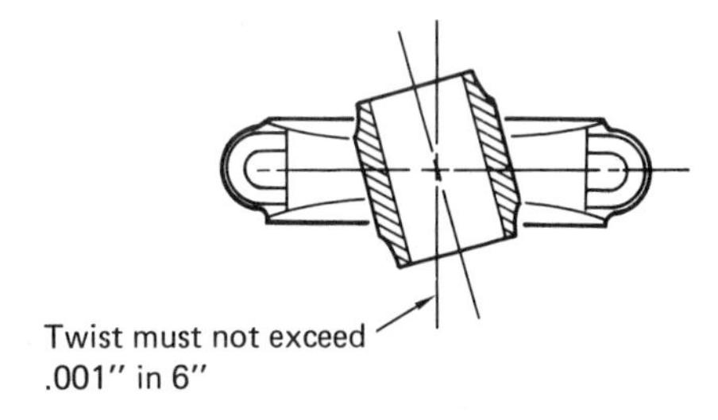


*FIGURE 10-46 Bend and twist conditions in connecting rods*

*FIGURE 10-47 Rod bearing wear caused by a bent connecting rod*

*FIGURE 10-48 A piston wear pattern associated with a bent connecting rod*

*FIGURE 10-49 Gauging and correcting bend in a K.O. Lee rod aligner*

abnormal wear and stress on connecting rod bearings (see Fig. 10-47), piston pins, and pistons (see Fig. 10-48).

Rod alignment is often done after the piston is assembled to the rod. Bend may be corrected with a notched pry bar (see Fig. 10-49). Similarly, twist may also be corrected with a notched pry bar (see Fig. 10-50).

Note, however, that twist is checked with

*FIGURE 10-50 Gauging and correcting twist in a K.O. Lee rod aligner*

*FIGURE 10-51 Bend being checked in reference to a full-floating piston pin*

*FIGURE 10-52 Twist being checked in reference to a full-floating piston pin*

the piston turned to one side. Both of these checks and corrections may be made in reference to the piston pin for full-floating designs (see Figs. 10-51 and 10-52).

There is a third condition that is occasionally found. The third condition is called *offset* and may be part of the design or the result of an incorrectly straightened rod. As a part of engine design, rod offset may be used to position the small end of the rod on the centerline of the cylinder (see Fig. 10-53). A connecting rod can also be made offset incorrectly by bending at the wrong point when correcting for bend.

It is important that rods designed with an offset be installed in the correct position. Installing one of these rods in a reverse position can cause engine noise and wear as a result of interference with the piston pin bosses and the crankshaft (see Fig. 10-54). An incorrectly offset rod can cause the same problems.

Rod offset can be checked by measuring the distance between the rod and the face of

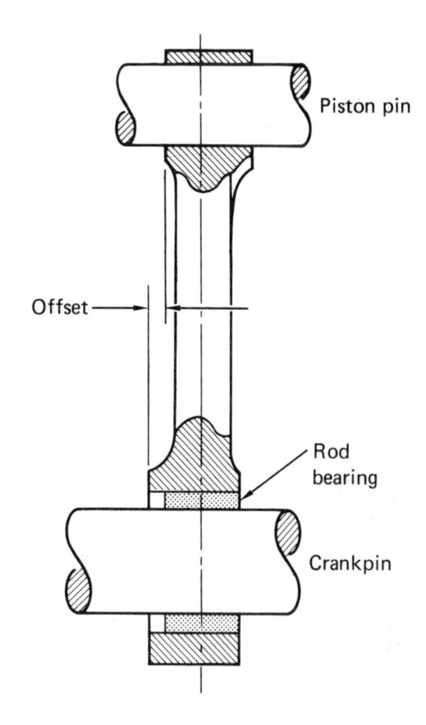


*FIGURE 10-53 A connecting rod designed with offset.*

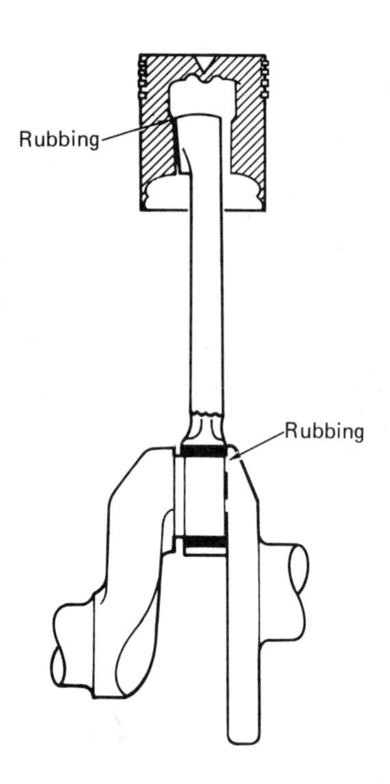


*FIGURE 10-54 Problems caused by the incorrect offset of connecting rods*

*FIGURE 10-55 Checking for offset in a Sunnen rod aligner*

the rod aligner (see Fig. 10-55). The measurement is taken once and then repeated after reversing the position of the connecting rod in the aligner. The difference in the two measurements is the amount of offset.

*FIGURE 10-56 Straightening a camshaft*

## REGRINDING THE CAMSHAFT AND VALVE LIFTERS

If it has been determined that the camshaft is worn beyond useable limits, the shaft must be replaced or salvaged by regrinding. Before regrinding, camshaft lobes are inspected to determine the extent of wear. If badly worn, individual cam lobes may be built up by welding.

The regrinding process is begun by checking the camshaft for straightness and straightening it as required (see Fig. 10-56). Each cam

*FIGURE 10-57 Grinding a camshaft*

*FIGURE 10-58 Applying a scuff-resistant coating*

lobe is then reground, usually to original equipment specifications (see Fig. 10-57).

The camshaft is treated with a scuff-resistant coating (see Fig. 10-58) following regrinding. The scuff-resistant properties of the camshaft surface add to the service life of the camshaft and are especially valuable during break-in, when wear patterns are established. However, this same coating is removed from cam journals by polishing (see Fig. 10-59). The cam journals run in relatively soft bearings, and a highly polished surface finish is more desirable.

*FIGURE 10-59 Polishing camshaft journals*

As mentioned, camshafts are usually ground to original equipment specifications. One common question is "How is specified lift maintained?" Keep in mind that each cam lobe is ground around its full circumference (see Fig. 10-60). Lift is also determined by the *difference* in distances from the nose of the cam lobe to the center of rotation and from the heel (on the base circle) to the center of

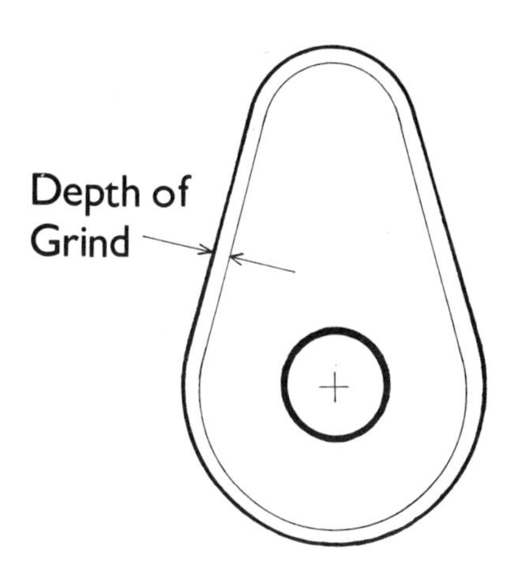


*FIGURE 10-60 A cam lobe is reduced in overall size by regrinding*

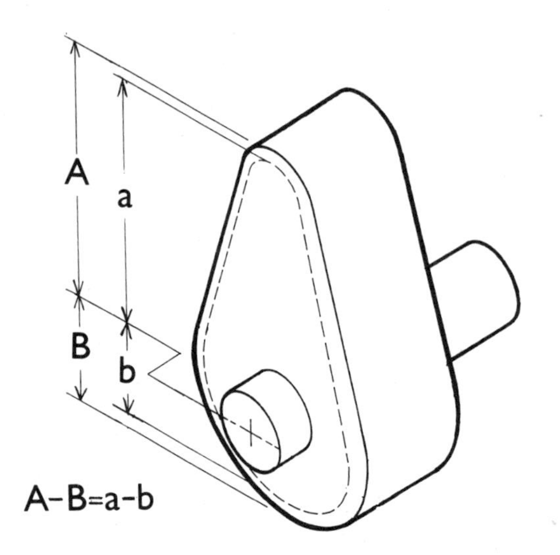


*FIGURE 10-61 Dimension A minus B is equal to dimension a minus b. Because of this, cam lift remains the same after grinding.*

*FIGURE 10-62 Regrinding a valve lifter*

*FIGURE 10-63 Valve lifters before and after regrinding*

rotation. The *difference* in these distances remains the same after grinding (see Fig. 10-61).

As mentioned earlier, the bases of most valve lifters are crowned slightly. This configuration can also be restored by regrinding. However, it is common practice to limit this process to solid, nonhydraulic valve lifters. The process requires a specialized tappet grinder. The rotation of the valve lifter against the contoured face of the grinding wheel restores the intended configuration to the valve lifter base (see Figs. 10-62 and 10-63).

## CRANKSHAFT GRINDING AND POLISHING

Crankshafts are often worn or damaged to the extent that regrinding is required. As mentioned earlier, wear may be apparent as a taper, out of roundness, or a scored surface finish. There are other less apparent conditions that may be disclosed during crankshaft reconditioning procedures. For example, shafts are occasionally found to be cracked or bent.

The crankshafts are cleaned before inspection and regrinding. Cleaning is often done by ordinary hot tanking methods, but particular care is given to handling shafts so that further surface damage or bending is prevented. Crankshafts are then inspected for cracks or breaks by means of magnetic particle inspection. The shaft is first sprayed with a fluid containing the magnetic particles. It is then magnetized, and cracks, if any, are viewed under a black light (see Fig. 10-64).

*FIGURE 10-64 A crankshaft in the wet maganflux machine*

Before they are reground, crankshafts are checked for straightness in V-blocks with a dial indicator (see Fig. 10-65). Automotive crankshafts can be straightened in the V-blocks with a punch similar to a chisel with a rounded edge. The V-blocks are positioned to each side of the bend, and the shaft is straightened by rapping the fillets of the journal with the punch (see Fig. 10-66). Using a punch in this manner stabilizes the shaft in the straight position and keeps it from springing back to the bent condition. It may be necessary to shift the V-blocks and strike the fillets of adjacent journals or crankpins to obtain the required straightness.

*FIGURE 10-65 Checking crankshaft straightness in V-blocks*

*FIGURE 10-66 Straightening a bent crankshaft*

*FIGURE 10-67 The appearance of a damaged rod throw built up by welding*

Badly damaged journals and crankpins may be built up by welding. The welding is often done by "submerged arc" welding machines. The process involves covering the welding arc with flux powder so that the weld metal remains clean, free of porosity, and comparable in hardness to the original shaft (see Fig. 10-67). Building up by welding makes it possible to grind shafts a minimum amount undersize.

Hardened crankshafts that have been overheated as a result of bearing failure are tested to see if they have softened (see Figs. 10-68 and 10-69). Hardened shafts that are damaged and must be reground may be rehardened using a nitride hardening process as at the factory. However, facilities for doing this work are not readily available in all areas of the country. As a result, many crankshafts are ground small, plated with hard chrome, and reground to size. They may also be given a very thin (about .0003 inch, or .008 mm) "flash" chrome treatment and polished.

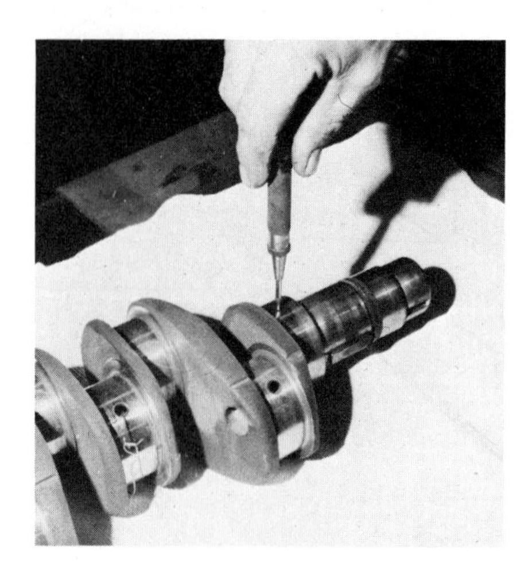

*FIGURE 10-68 Making an impression on the crankshaft with a penetrator as the first step in hardness testing*

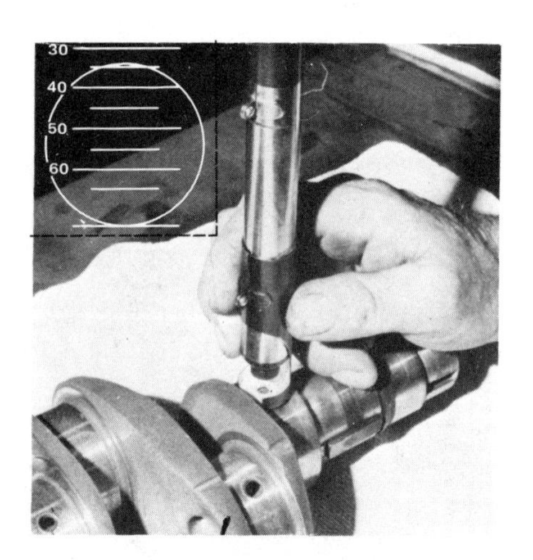


*FIGURE 10-69 Checking the impression left by the penetrant to evaluate the hardness*

Crankshafts are commonly ground to undersizes of .010 inch (.25 mm), .020 inch (.50 mm), and .030 inch (.75 mm). Occasionally, shafts may be ground as much as .060 inch (1.52 mm) undersize, although crankshaft welding has reduced the need for such undersizes.

It is standard practice to grind all main journals to the same undersize. For one thing, it would be extremely difficult, if not impossible, for the machinist to maintain a straight centerline without grinding all journals in the same machine setup.

The same practice is followed for rod journals (also referred to as crankpins or throws), although the undersize may vary from the main journals. For example, the main journals may be .010 inch (.25 mm) undersize, and the rod journals may be ground .020 inch (.50 mm) undersize. While each of the crankpins could be ground to varying undersizes, it is not generally done.

Crankshaft grinding machines are specially designed for the purpose (see Fig. 10-70). They are expensive and, therefore, sometimes limited to production engine rebuilders or a few well-equipped automotive machine shops. The operator is usually a skilled specialist who must continually make checks and adjustments so that taper, roundness, size, fillet radius, and surface finish are all held within limits.

The final surface finish quality necessary for normal bearing service is obtained by polishing after grinding. Grinding leaves microscopic burrs that are best removed by a fine grit polishing belt (see Fig. 10-71). Thrust surfaces and seal surfaces are also polished. Polishing by this procedure is recommended for engine overhauls, even when regrinding is not done, to remove minor scoring and marks caused by measuring and handling. The amount of metal removed is usually less than .0002 inch (.005 mm) on the diameter.

*FIGURE 10-70 A machine for regrinding a crankshaft*

*FIGURE 10-71 Polishing a crankshaft*

## FLYWHEEL RESURFACING

Resurfacing the flywheel on engines equipped with standard transmissions is an operation that is frequently performed as a part of clutch service. It is also an obvious convenience to service the clutch and flywheel during engine overhaul or rebuilding. The resurfacing operation restores flatness, removes heat checking (surface cracks) produced by clutch slippage, and aids in restoring smooth clutch operation. The resurfacing may be done in flywheel grinders (see Fig. 10-72), other automotive resurfacing machines, or brake drum lathes with adapters.

*FIGURE 10-72 A Winona flywheel grinding machine*

Be sure to magnaflux flywheel surfaces because clutch slippage can overheat the flywheel and cause cracking. Some cracks are superficial and can be removed by resurfacing. However, some may penetrate the flywheel (see Fig. 10-73). A flywheel failure is hazardous to the driver, passengers, or anyone near the engine.

It is important that the clutch surface be machined flat and parallel to the side of the flywheel that bolts to the crankshaft. Maintaining this relationship should keep flywheel run-out (wobble at the face) within .005 inch (.13 mm) (see Fig. 10-74).

*FIGURE 10-73 Magnetic powder is collected along cracks on flywheel surface*

*FIGURE 10-74 Checking flywheel run-out with a dial indicator*

Limits for the amount of material that may be removed from the clutch surface are often not specified. Where given, specifications usually permit as much as .030 inch to .035 inch (.75 mm to .89 mm) of resurfacing. Recessed clutch surfaces require machining in two steps. The same depth of machining required for cleanup of the recessed clutch surface should be made at the outside surface where the pressure plate is attached. This two step procedure maintains normal pressure for the clutch pressure plate.

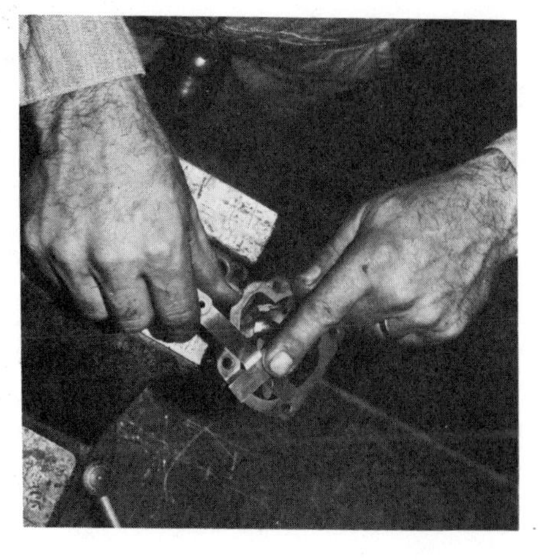

*FIGURE 10-75 Checking gear end clearance in an oil pump*

## OIL PUMP SERVICE

Many rebuilders just replace oil pumps with new or rebuilt units. Some mechanics may check the pump and reuse it if it is operative in an acceptable manner. There are inspection procedures that may be used to determine whether pumps are within the specified service limits. It is recommended that these inspections be made so that any decision regarding the repair or replacement of the pump may be made on the basis of specifications.

Gear pumps are checked for end clearance of the gears in the pump housing (see Fig. 10-75) and for clearance between the gear teeth and the housing (see Fig. 10-76). Approximate limits are .003 inch (.05 mm to .10 mm) for end clearance and .005 inch (.10 mm to .15 mm) for gear-to-housing clearance.

Rotor pumps are checked for end clearance in the same manner as gear pumps. Clearance between inside and outside rotors is checked with a feeler gauge (see Fig. 10-77). Clearance between the outer rotor and the pump housing is also checked with a feeler gauge (see Fig. 10-78). Approximate limits are a maximum of .010 inch (.25 mm) between inside and outside rotors and .014 inch (.36 mm) between outside rotor and housing.

*FIGURE 10-76 Checking clearance between gear teeth and housing in an oil pump*

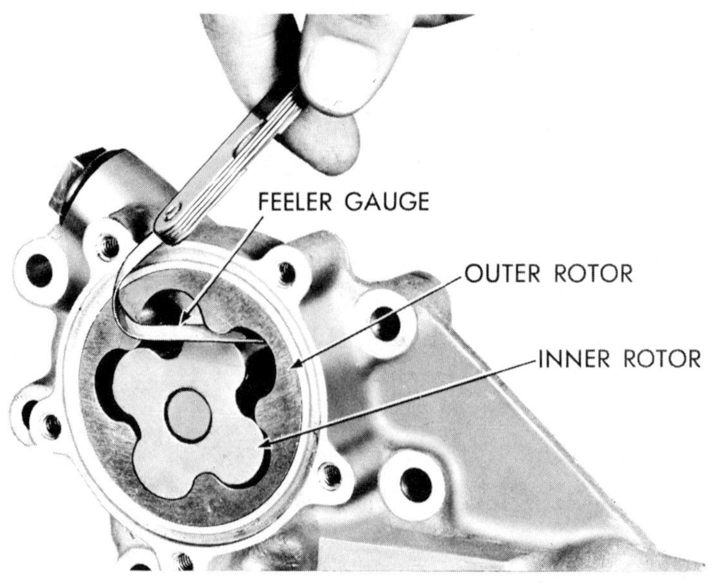


*FIGURE 10-77 Checking clearance between rotors in an oil pump (Courtesy of Chrysler Corp.)*

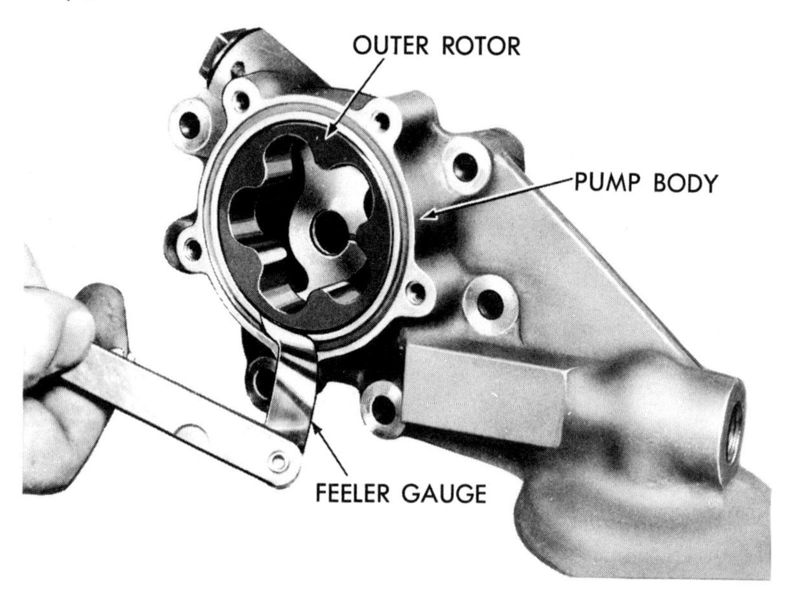


*FIGURE 10-78 Checking outer rotor to housing clearance in an oil pump (Courtesy of Chrysler Corp.)*

The end covers for both gear and rotor pumps are checked for wear. Small amounts of wear may be removed by hand lapping on abrasive paper. Oil pump drives for both types of pumps are also checked for wear. This is particularly true for hexagonal drives that are in position between the distributor and the oil pump. Watch for a visible rounding of the ends of these drives (see Fig. 10-79).

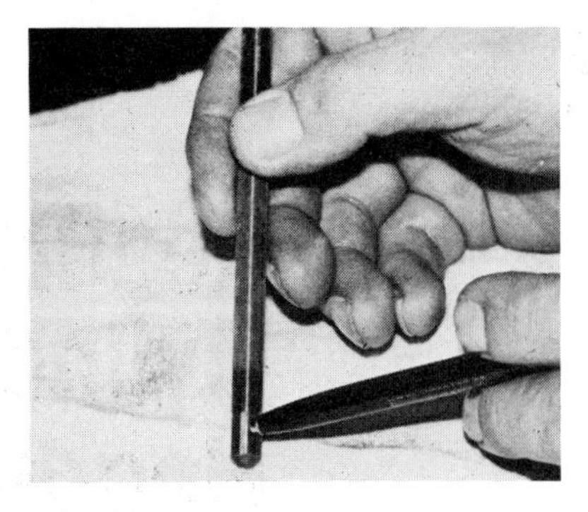

*FIGURE 10-79 Watch for rounding of hexagonal oil pump drives*

Repair kits are available with gears or rotors, gaskets or seals, and pressure-relief springs. Oil pump drives may be replaced separately. Pumps are reconditioned by giving the pump a thorough cleaning and by replacing worn parts as needed.

Oil pump pickups are attached to the oil pump body, or possibly to a passage in the engine block leading to the pump, with screws or by a press-fit (see Fig. 10-80). The pickup must be securely attached so that it does not vibrate loose. It must also be well sealed at the point of attachment so that air cannot be drawn into the pump. The seal may be obtained by a gasket, O-ring, or a good press-fit. A pickup that comes loose or permits oil to become aerated will cause engine failure. It is not uncommon to braze or solder press-fit pickups in place as a precaution. The internal parts, especially the pressure relief spring and valve, should be removed during brazing or soldering.

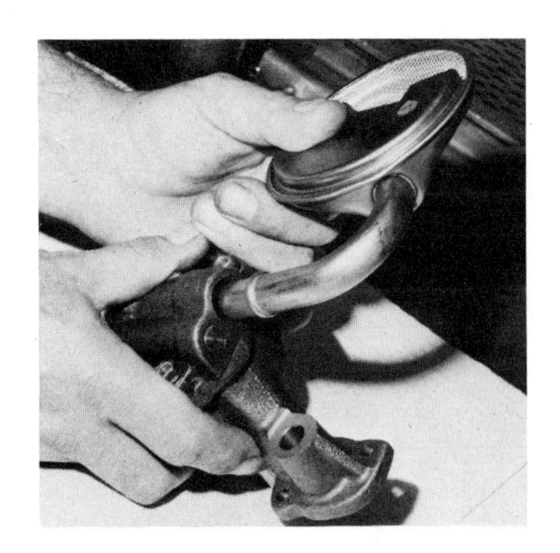

*FIGURE 10-80 A press-in type of oil pump pickup*

# 11

# Resurfacing Cylinder Heads and Blocks

The primary reason for resurfacing is to ensure head gasket sealing. Surfaces on cylinder heads and blocks become warped because of overheating or improperly tightened head bolts. They are also damaged by the flow of gases through leaking head gaskets. Resurfacing restores flatness and improves surface finishes.

Resurfacing is done routinely on rebuilt or remanufactured engines because they are being restored to new specifications and will be expected to operate as long as the original. During overhauls or valve grinding, surfaces are remachined only as required. Inspections are made for flatness, dents or scratches, and corrosion around water passages (especially on aluminum heads). Probably half of the cylinder heads checked will require resurfacing.

Be especially careful to check surfaces if a head gasket failure has occurred. It is recommended that both heads on V-block engines be resurfaced even if only one head gasket failed because both heads were subjected to the same operating conditions. This practice also keeps compression equal on both sides. Be sure to check the engine block surface for damage from gas flow through the leak in the gasket.

## RESURFACING MACHINES

*FIGURE 11-1 A Van Norman 530 mini-broach resurfacing machine*

Some machines are suited to quick, routine resurfacing operations on cylinder heads (see Fig. 11-1). The machines may remove metal by grinding or by milling. Some machines will resurface engine blocks as well as cylinder heads.

Grinding machines produce better surface finishes than milling machines (see Figs. 11-2 and 11-3). Also, hard spots in castings are easily machined by grinding. However, the casting and grinding wheel are deflected away from each other slightly during grinding. Because of this deflection, the depth of each cut is small and extra passes, without any increase in the depth of cut, are required to obtain a flat surface. These extra passes are referred to as "sparking out." The small depth of cuts and the sparking out cause grinding to be somewhat slower than milling.

In milling machines, it is common to make one deep cut to clean up the casting and one shallow cut to improve flatness and the surface finish (see Figs. 11-4 and 11-5). However, hard spots in castings will dull the milling cutters and require that they be resharpened.

*FIGURE 11-2 A Kwik-Way wet grinder resurfacing a block*

*FIGURE 11-3 A Kwik-Way wet grinder being set up to resurface a head*

*FIGURE 11-4 A Storm Vulcan Block-Master resurfacing a cylinder head*

*FIGURE 11-5 A Storm Vulcan Block-Master resurfacing a block*

will dull the milling cutters and require that they be resharpened.

## PRECAUTIONS TO FOLLOW

As a rule, remove as little as possible to restore flatness. An inspection with a straightedge and a feeler gauge has limited accuracy in determining how much is required for cleanup. For example, a .005-inch (.13 mm) feeler gauge might just slip under the straightedge, but it may require an .008-inch (.20 mm) cut to cleanup the surface. Most surfaces will cleanup within .010 inch (.25 mm).

The hazards involved in removing excess amounts are numerous. The most obvious is the change in compression ratio. Increases in the compression ratio will be slight in most cases and, therefore, cause no problem. However, it is possible for the increase in compression to require switching to high octane fuel or retarding the ignition timing. Neither alternative is desirable. Consider that many cars are now equipped with catalytic converters. A change to high octane (leaded) fuel is impossible, and the required changes in timing will reduce power and economy.

In some engines, valve-to-piston clearance is minimal. It is possible to resurface a cylinder head and cause interference between valves and pistons (see Fig. 11-6). This can occur regardless of whether the engine is a push rod (overhead valve) or overhead cam design.

In the case of push rod engines with non-adjustable rocker arms, there is another change to consider. While resurfacing brings the cylinder head and block closer together, the push rod length remains the same. The result is that the push rods open the valves because the head is closer to the block. This is a situation similar to having valve stems too long. The result will be at least a loss of compression caused by open valves, and possibly bent valves.

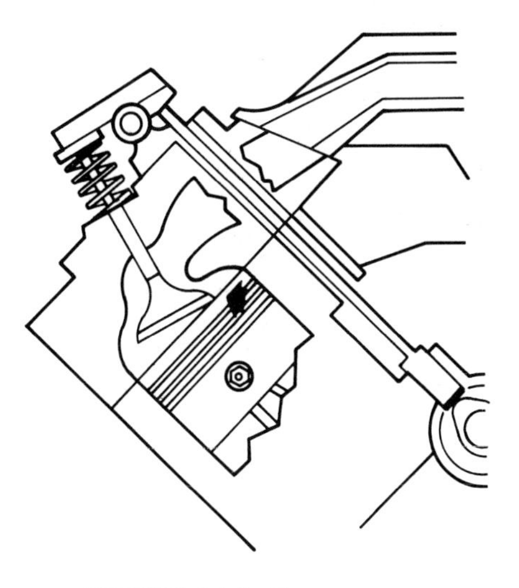

*FIGURE 11-6 Excess resurfacing can cause interference between pistons and valves. (Courtesy of Fel-Pro Inc.)*

Resurfacing causes valve timing changes in overhead camshaft engines because of timing chain (or belt) slack created as the camshaft and crankshaft move closer together. Consider that slack on the driving side of a timing chain is taken up by crankshaft rotation and slack on the opposite side of the chain is taken up by a chain tensioner. Valve timing retards (valve timing is late) because the camshaft remains stationary until the crankshaft rotation takes up the chain slack. In fact, valve timing will retard approximately 1° for every 0.020 inch (.51 mm) removed. Resurfacing should be kept to a minimum, certainly within any specified limits, because retarded valve timing reduces engine torque at low RPM.

Another problem is the change in intake manifold alignment on V-block engines. On these engines, resurfacing not only causes the heads to move closer to the block, but also causes them to move closer together. Unfor-

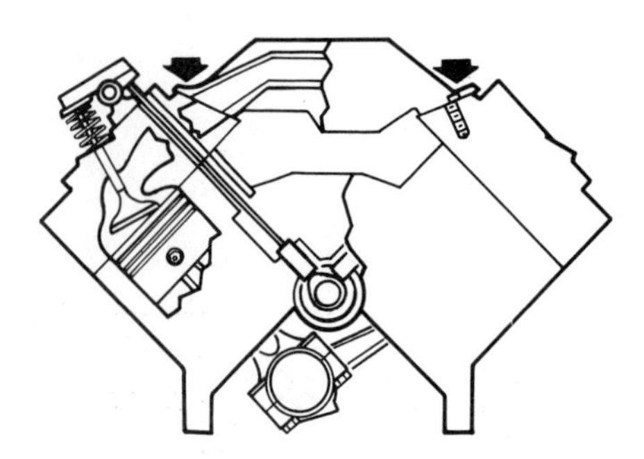

*FIGURE 11-7 Misalignment between cylinder heads and the intake manifold caused by excess resurfacing (Courtesy of Fel-Pro Inc.)*

*FIGURE 11-8 Thicker head gaskets used to compensate for resurfacing (Courtesy of Fel-Pro Inc.)*

tunately, the intake manifold remains just as wide and must be positioned between the cylinder heads. The result is that manifold bolt holes and intake ports move out of alignment (see Fig. 11-7). The most apparent problem is the possibility of vacuum leaks. Less apparent is the oil pull-over into intake ports resulting from a poor seal between the intake manifold gasket and the interior of the engine.

Keep in mind that this may not be the first time these heads or blocks have been resurfaced. The amount resurfaced now will be in addition to any resurfacing done before. Try to limit the total amount of resurfacing of heads and blocks combined to approximately .015 inch (.38 mm). This amount can be compensated for by using thicker head gaskets (see Fig. 11-8). Check with a parts supply store, and specify the thicker gaskets for engine assembly after resurfacing.

## CORRECTING INTAKE MANIFOLD ALIGNMENT

Some cylinder heads warp so badly that they can be salvaged only by resurfacing in amounts more than what has been recommended. Extreme amounts of resurfacing are routinely done on high performance engines. For example, combustion chamber volumes are reduced to the minimum by means of resurfacing. On the same engine, block surfaces are machined to reduce the distance from the block surface to the pistons (called deck height) to the absolute minimum. About the only limitation to resurfacing for high performance is valve-to-piston clearance and the fuel to be used.

In such cases of extreme resurfacing, manifold alignment can be corrected only by a series of additional machining operations.

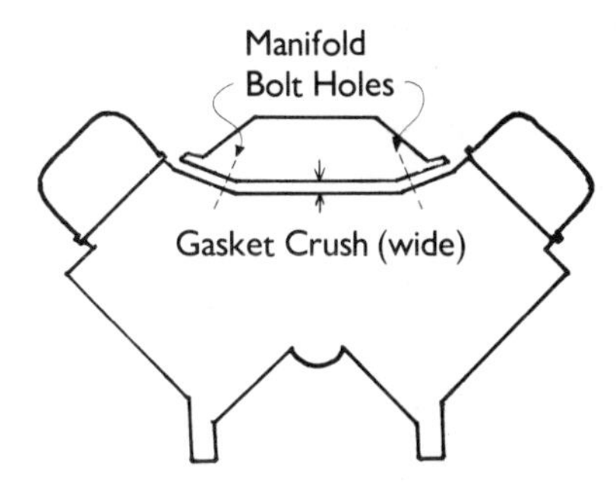


*FIGURE 11-9 The normal alignment of the manifold, cylinder heads, and V-block*

*FIGURE 11-10 Resurfacing the intake sides of V-block cylinder heads in a Storm Vulcan machine*

First, study an engine assembly before resurfacing. Manifold bolts and intake ports are in alignment. Note also that there is a certain amount of "crush," or gasket clearance, between the ends of the manifold and the top of the block (see Fig. 11-9). The alignment of bolts and ports and the crush at the ends of the manifold must be restored to this condition after all of the corrections are made.

The first correction is to restore the alignment of bolts and ports. This is done by resurfacing the intake side of each cylinder head (see Fig. 11-10). This resurfacing increases the distance between cylinder heads so that the manifold will fit again. Note, however, that the amount of clearance (gasket crush) at the ends of the manifold is reduced (see Fig. 11-11), because machining the intake sides of the cylinder heads causes the manifold to sit lower on the engine. The second correction is to restore the original crush by resurfacing the top of the block beneath the manifold (see Fig. 11-12).

Keep in mind also that if the distributor is installed through the manifold or on top of the block where resurfacing was done, it too will sit lower. And, if the distributor drives the oil pump, the depth of engagement of the distributor and oil pump will be increased.

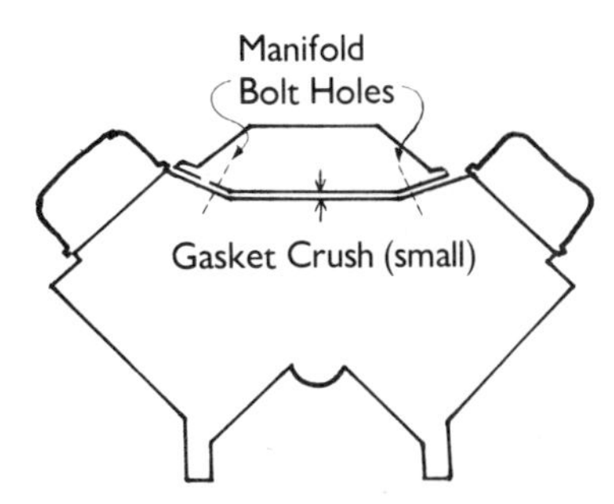


*FIGURE 11-11 Port and manifold bolt alignment has now been corrected. Note, however, that the gasket crush is excessive.*

*FIGURE 11-12 Setting up to machine the top of the block to restore the correct gasket crush*

*FIGURE 11-13 An exploded view of the relationship between the distributor, camshaft, and oil pump drive*

The depth of engagement of the distributor drive gear with the camshaft will also be changed. All of these components assemble in such a way that any changes to the top of the block or to the elevation of the manifold affect other parts (see Fig. 11-13). Alignments can be completely restored by shimming the distributor by an amount equal to the resurfacing of the block (under the manifold).

There is a set of stock removal ratios for each engine to determine the amount of resurfacing in each step. For example, based on the amount removed from the heads and block, the ratios required for a 350-cubic-inch Chevrolet are 1.23:1 (intake side of heads) and 1.71:1 (top of the block). These ratios are used as follows:

If the amount removed from cylinder heads and block to restore flatness is

0.010 in. (.25 mm),

the amount to remove from the intake side of cylinder heads to align bolts and ports is

$$0.010 \times 1.2 = 0.012 \text{ in. } (.30 \text{ mm})$$

and the amount to remove from the top of the block to restore gasket clearance is

$$0.010 \times 1.7 = 0.017 \text{ in. } (.43 \text{ mm})$$

The ratios to be used are different for each type of engine. The difference is related to the angle between the head gasket surface and the intake side of cylinder heads (see Fig. 11-14). The ratios also depend on the angle between the head gasket side of the block and the top surface beneath the intake manifold. Fortunately, there are only two basic angles for engine blocks in use (see Fig. 11-15). A table of ratios for each of the two types of blocks is given below:

| Included Head Angle | *Complimentary Head Angle | Intake Side Ratio | Top of Block Ratio |
|---|---|---|---|
| **90° V-BLOCKS** | | | |
| 90° | 0° | 1.0 | 1.4 |
| 80° | 10° | 1.2 | 1.7 |
| 75° | 15° | 1.4 | 2.0 |
| 70° | 20° | 1.7 | 2.3 |
| **60° V-BLOCKS** | | | |
| 90° | 0° | 0.6 | 1.2 |
| 80° | 10° | 0.7 | 1.3 |
| 70° | 20° | 0.8 | 1.5 |

*90° minus the included head angle

It must be pointed out that there are some options to the machining steps described. For example, the sides of the intake manifold can be machined instead of the sides of the heads. The same ratios would be used to determine the amount of resurfacing. Similarly, the underside of the manifold could be machined instead of the top of the block. Again, the same ratios as used for the top of the block would apply.

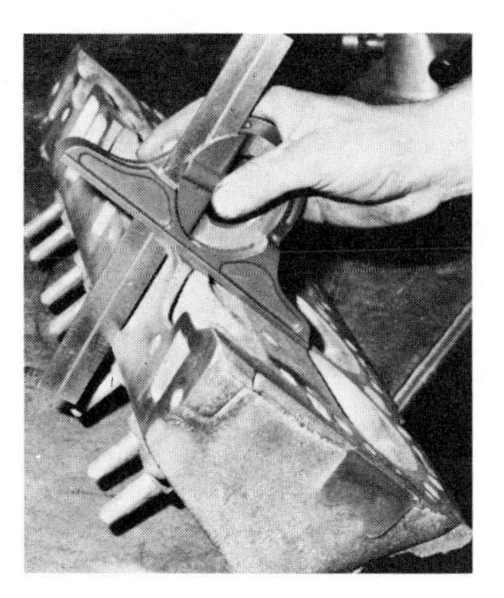

*FIGURE 11-14 Measuring the angle of a cylinder head with a protractor*

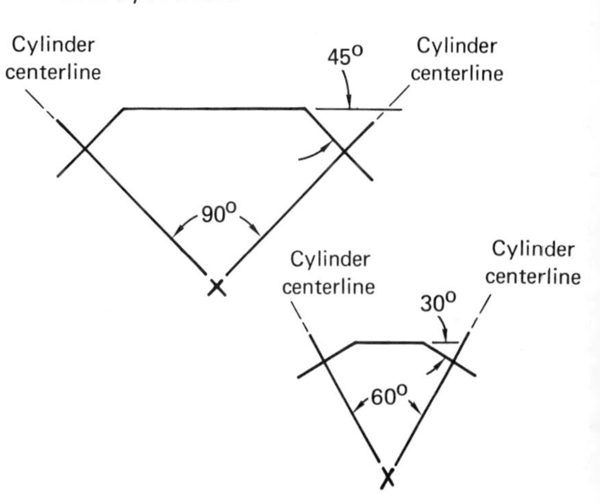


*FIGURE 11-15 V-blocks with 60° and 90° cylinder arrangements*

## DETERMINING RATIOS

As shown, stock removal ratios vary according to cylinder head and block angles. These ratios can be derived from a combination of mathematical calculations. Performing such calculations requires some mathematics background and does not necessarily make the problems in engine assembly any easier to understand.

Another method of determining the ratios is by graphic solution. A reconstruction of the problem can be easily drawn with only a few lines. From such a drawing, all ratios can be determined by measuring the sides of a triangle.

Using a 350-cubic-inch Chevrolet as an example, begin by making a drawing showing the intersection of one side of the block, the top of the block, and one side of the cylinder head (see Fig. 11-16). Take the angles directly from the engine.

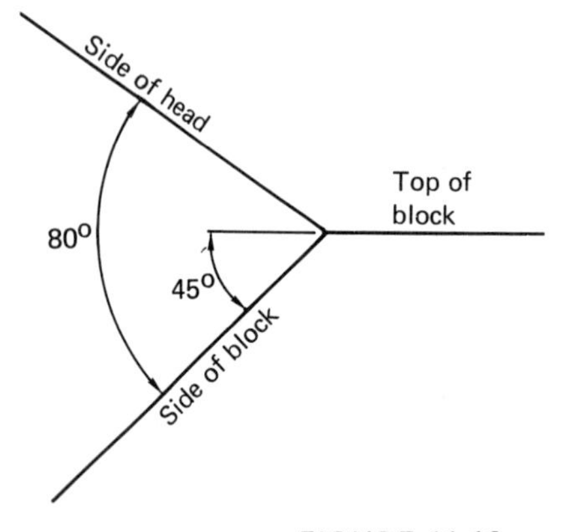


*FIGURE 11-16*

The next step is to establish a reference point on the intake side of the cylinder head. It may help to imagine this point as being the center of a manifold bolt hole (see Fig. 11-17).

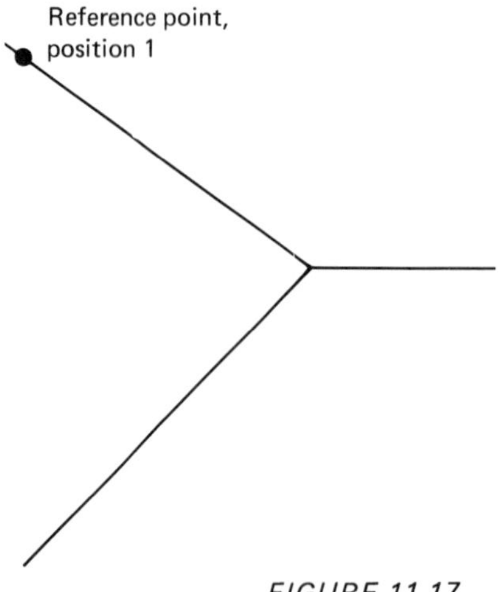


*FIGURE 11-17*

Next, consider that this point will move when the cylinder head or block is resurfaced. To be exact, it will move at a 90° angle to the side of the block (see Fig. 11-18).

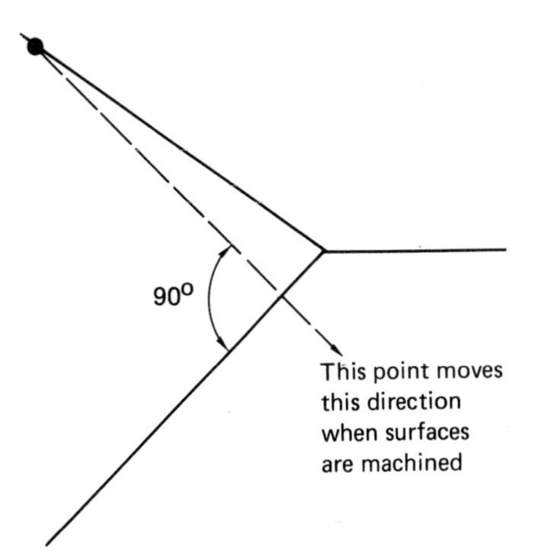


*FIGURE 11-18*

For purposes of determining ratios, imagine that 1 inch has been removed by resurfacing. Measure along the direction of movement 1 inch from the reference point and draw another point (see Fig. 11-19).

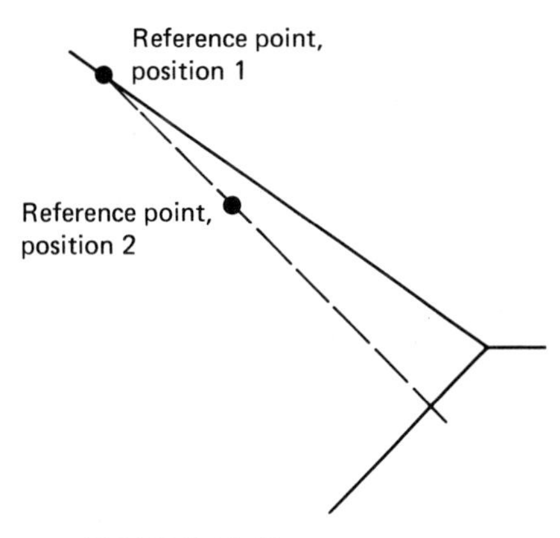


*FIGURE 11-19*

Position 2 of the reference point indicates that the cylinder head has moved toward the center of the engine block. The reference point must be moved outward again so that the manifold will fit between the cylinder heads. Resurfacing the intake side of the head will move the reference point outward along a line 90° to the intake side of the head (see Fig. 11-20).

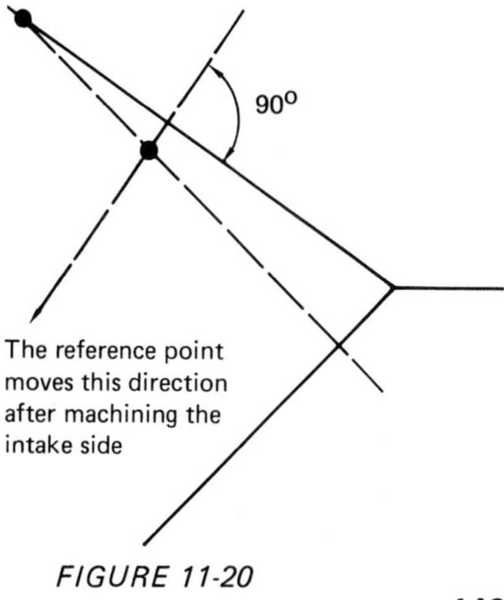


*FIGURE 11-20*

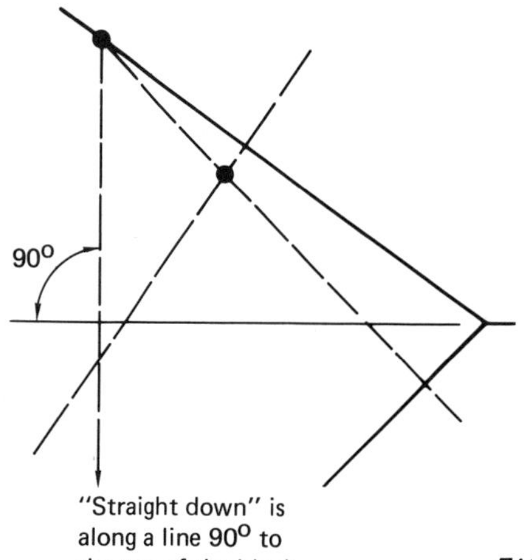


Resurfacing the intake side must move the reference point outward until it is directly under its original position. Draw a line straight down (at a 90° angle to the top of the block) from the original position of the reference point (see Fig. 11-21).

*FIGURE 11-21*

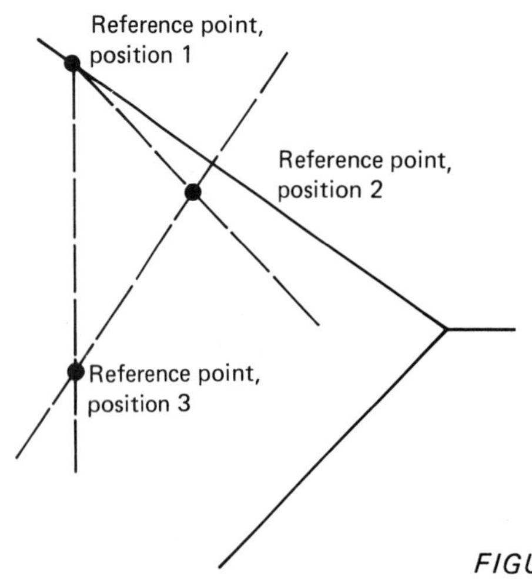


Reference point position 3 is at the intersection of the last two lines drawn. This point completes the triangle required to determine ratios (see Fig. 11-22).

*FIGURE 11-22*

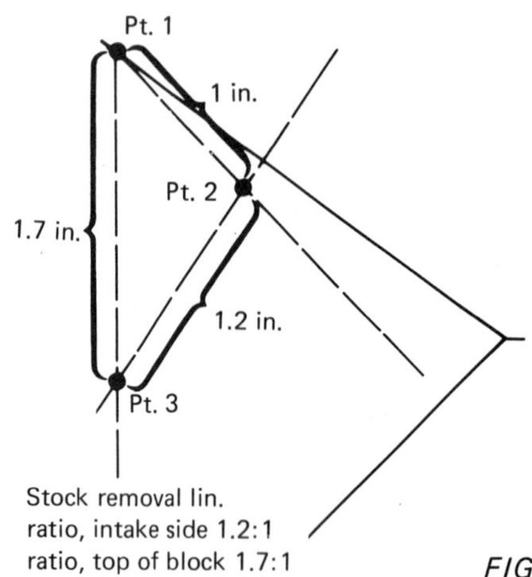


The next step is to measure the sides of the triangle. The measured distances will be equivalent to the stock removal ratios (see Fig. 11-23).

*FIGURE 11-23*

In the case described, the intake side ratio is 1.2:1 and the top of block ratio is 1.7:1. If 1 inch were actually removed from the face of a cylinder head, the reference point would move from position 1 to position 2. For correct manifold bolt and port alignment, the intake side of the head must be machined until the reference point moves from position 2 to position 3. For correct clearance under the manifold for a gasket, the top of the block must be machined an amount equal to the distance between reference point position 1 and position 3.

This method of determining ratios can be used for any V-block engine. Just measure the angle from the top of the block to the side of the block. Also measure the included angle of the cylinder head. Then repeat the drawing steps as described.

# 12

# Engine Balancing

Engine balancing is a procedure that *minimizes* engine vibrations. It does not necessarily *eliminate* vibration because balance also depends on inherent design characteristics associated with the number of cylinders and the type of crankshaft and engine block.

Balancing is especially important in high speed engines because the forces that cause vibration multiply as engine speed increases (see Fig. 12-1). Racing engines, therefore, are always balanced. Heavy-duty truck engines may also be balanced to reduce the unbalance forces acting on the crankshaft. Passenger car engines may be rebalanced at the option of the owner to make engine operation smoother than normally expected.

Balancing is recommended whenever engine parts are mismatched. This means engine parts that have come from different sources.

| RPM | Oz. | Gms. |
|---|---|---|
| 500 | 7.3 | 207 |
| 1000 | 19 | 539 |
| 2000 | 117 | 3,317 |
| 3000 | 263 | 7,456 |
| 4000 | 464 | 13,154 |
| 5000 | 720 | 20,412 |

Note: Values based upon a one ounce weight placed one inch from center.

*FIGURE 12-1 As rotational speed (velocity) increases, the effect of force increases dramatically*

The pistons may come from a supplier of replacement parts. Some rods may be original, and others salvaged from another engine. The crankshaft may have been exchanged for another. The result as far as balance is concerned is anyone's guess.

## CORRECTING RECIPROCATING WEIGHT

All weight which travels up and down in the cylinder is considered reciprocating weight. This includes pistons, piston pins, and the small ends of connecting rods. Even piston rings and piston pin retainers for full-floating pins would be included. The total of all reciprocating weight in each cylinder should be made equal to reduce variations in force.

Correcting variations in reciprocating weight is the first step in engine balance. The pistons are weighed to find the lightest piston (see Fig. 12-2). All heavier pistons are then lightened to match the weight of the lightest piston. This is typically done by facing off the balance pads under the piston pin bosses (see Fig. 12-3). The piston pins may be included

*FIGURE 12-2 Comparing piston weights*

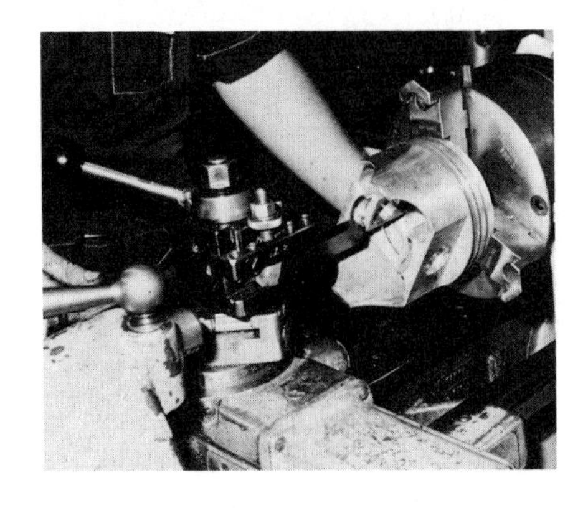

*FIGURE 12-3 Facing balance pads on a piston to remove weight*

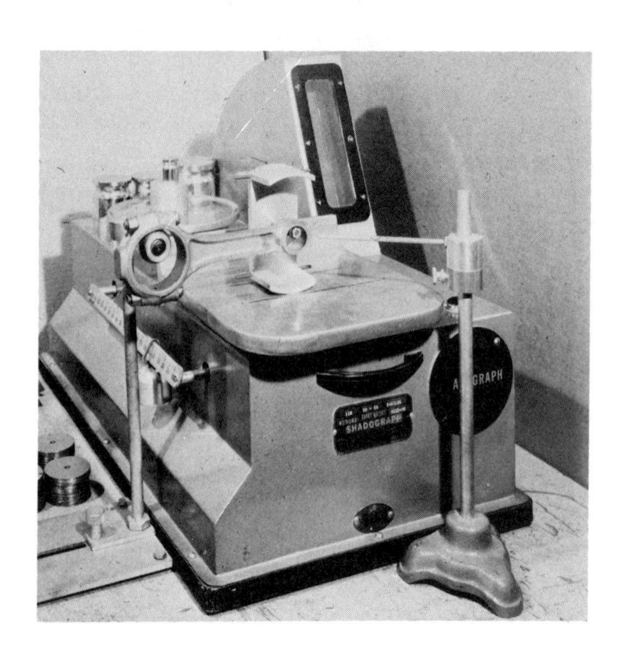


*FIGURE 12-4 Weighing the light ends (reciprocating ends) of connecting rods*

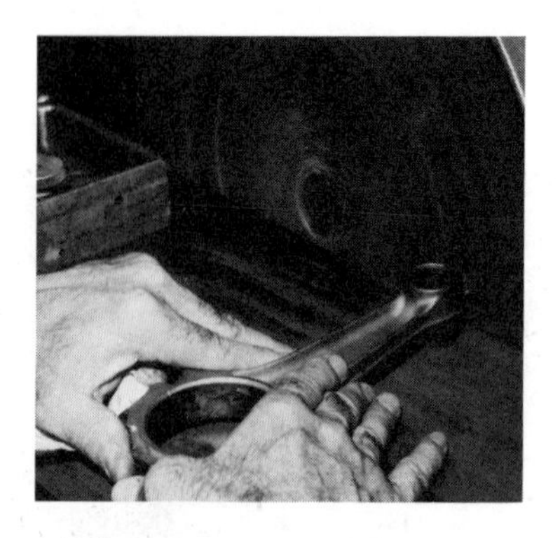

*FIGURE 12-5 Removing weight from connecting rod balance bosses to equalize light end weights*

in the weight of each piston or weighed separately. Correction is made to within 1/2 to one gram, depending on the range of engine speed expected. The piston rings and pin retainers, if used, are not corrected but their weights are included as a part of the total for reciprocating weight.

The small end of each connecting rod is weighed until the lightest is found (see Fig. 12-4). As with pistons, the weight of the heavier rods is corrected to match the weight of the lightest rod. The weight is removed by sanding or grinding the balance boss at the top of the rod (see Fig. 12-5). Weight must be carefully removed from all around the small end of the rod if a balance boss is not present.

## CORRECTING ROTATING WEIGHT

The next step in the balancing procedure is to eliminate variations in the rotating weight of the connecting rod housing bore ends. Rotating weight includes rod bearings, although variations in weight are negligible and are not corrected.

Remember that the small ends of the rods have already been matched in weight; therefore, any variations in the total weight will be due to excess weight at the housing bore ends. All rods are weighed (see Fig. 12-6) to find the lightest rod. The heavier rods are then made to match by sanding or grinding on the balance boss at the housing bore ends (see Fig. 12-7). As with reciprocating weights, a spread of 1/2 to one gram is used.

*Removing weight at the housing bore end of a connecting rod will make the small end slightly heavier.* Check the setup for weighing the small end, and it can be seen that the housing bore end acts as a lever lifting up on the small end. Removing weight from the housing bore end, therefore, causes the weight at the small end to increase. It is required that small end weights be rechecked and corrected as required after correcting weights at the housing bore ends.

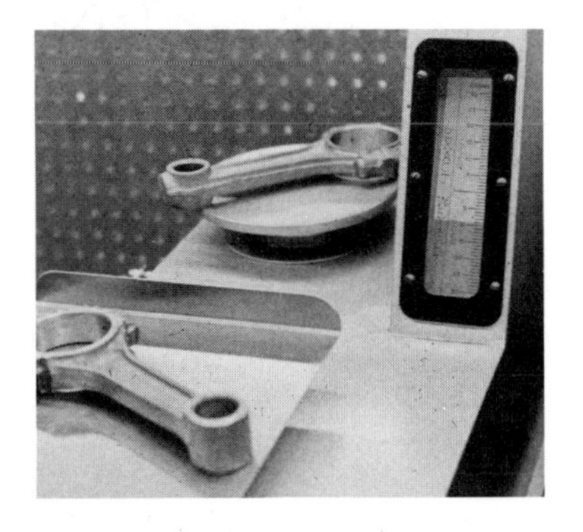

*FIGURE 12-6 Comparing total weights of connecting rods*

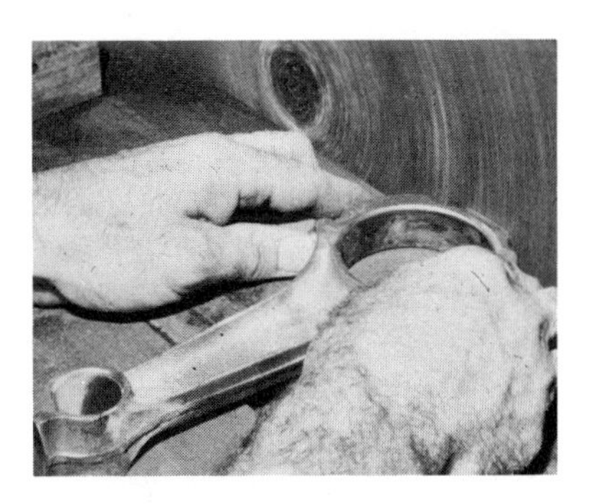

*FIGURE 12-7 Removing weight from balance bosses at the housing bore end to equalize connecting rod total weights*

## CRANKSHAFT BALANCING

A crankshaft may be *statically* balanced but not run smoothly in the engine. By static balance, it is meant that the weight is evenly distributed around the center of rotation. However, weight may be located on one side of center at one end and the other side of center at the opposite end of the shaft (see

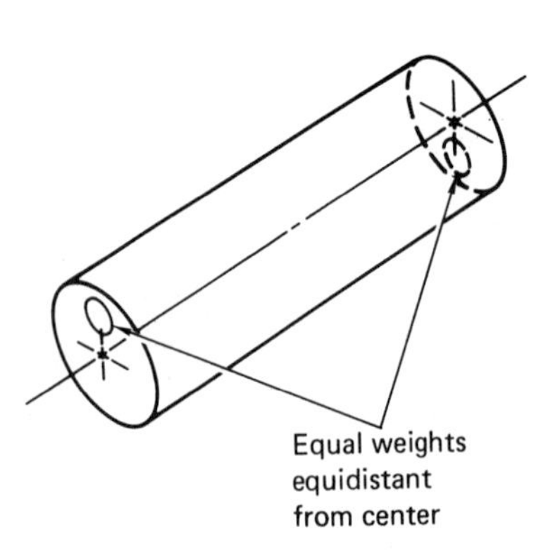


*FIGURE 12-8 A static-balanced cylindrical part. Weights are equal and equal distances from the center line*

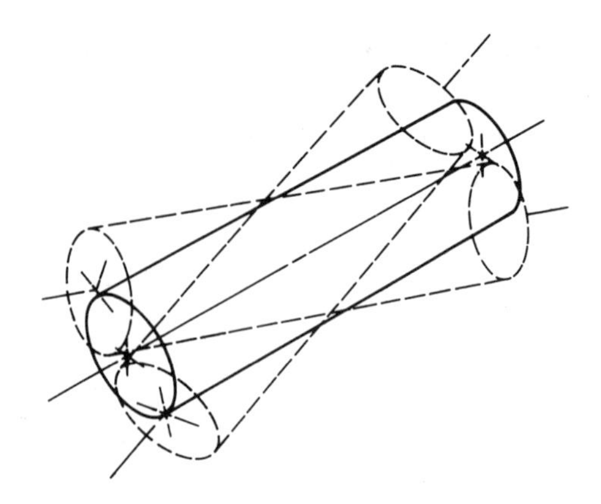

*FIGURE 12-9 The same static-balanced part develops a dynamic "wobble" when rotated. It is dynamically unbalanced because of the static balance weights at each end.*

Fig. 12-8). When the crankshaft is rotated it will wobble because *each end* of the crankshaft is unbalanced (see Fig. 12-9).

A crankshaft then is balanced in two *planes.* That is, unbalance is detected and corrected at each end of the crankshaft (see Fig. 12-10). The crankshaft is rotated (approximately 520 RPM) in the balancing machine, and weight is removed by drilling or added by

*FIGURE 12-10 A crankshaft with bob weights prepared for dynamic balancing. Note the dial indicators used to detect "wobble" (Courtesy of Bear Mfg. Co.)*

welding at one end at a time until balance at both ends is acceptable.

Flywheels and harmonic balancers should be checked for *external balance weights* before proceeding with crankshaft balancing. External weights are weighted sections on flywheels and harmonic balancers designed to be a part of crankshaft balance. If such weights are apparent, the crankshaft, flywheel, and harmonic balancer should be balanced as an assembly.

V-block engine crankshafts present a special problem in balancing. For example, the counterweights on V-8 shafts compensate for the crankpin plus the *rotating weight and half of the reciprocating weight.* This means that *bob weights* must be made up and added to the crankshaft for balancing (as shown in Fig. 12-10). Again, the bob weights are equal to all the rotating weight and half the reciprocating weight on each crankpin. As you can see, all balancing steps on pistons, pins, and connecting rods must be completed first.

Most V-block engines use the same formula, 100 percent rotating weight plus 50 percent reciprocating weight, for calculating bob weights. Exceptions to this formula are the General Motors 90° V-6 engines with *splayed crankpins.* The crankshafts for these engines have 3 crankpins with 2 connecting rods on each crankpin, just like a V-8. However, 1/2 of each crankpin is ground offset from the other 1/2 so that, in effect, there are 6 crankpins. The 3.8 Liter "even-firing" engine has a 30° offset between connecting rods on the same crankpin and bob weights are calculated by using 100 percent rotating weight plus 36.6 percent reciprocating weight. The 3.3 Liter odd-firing engine has an 18° offset between connecting rods on the same crankpin and bob weights are calculated by using

100 percent rotating weight plus 46 percent reciprocating weight.

In-line engine crankshafts usually do not require bob weights for balancing. The counterweights and crankpins on six-cylinder crankshafts are in pairs evenly spaced around the crankshaft center line and the forces cancel out. The same is true for the piston and rod assemblies because they are also evenly spaced around the crankshaft centerline. The counterweights and piston and rod assemblies only partly cancel unbalanced forces in four-cylinder crankshafts. While engine smoothness can be improved by balancing, some vertical forces of inertia will remain because of inherent design characteristics.

## BALANCING FLYWHEELS

The flywheel and any other part attached to the crankshaft may be balanced. This is done by adding the parts to the balanced crankshaft in the balancing machine. For example, the flywheel may be attached and balanced (see Fig. 12-11), and the clutch may be added to the flywheel and balanced (see Fig. 12-12). Harmonic balancers may be added to the opposite end of the shaft and balanced. It is

*FIGURE 12-11 The flywheel added to the crankshaft for balancing (Courtesy of Bear Mfg. Co.)*

*FIGURE 12-12 The clutch added to the flywheel for balancing (Courtesy of Bear Mfg. Co.)*

generally found that the clutch is most in need of balancing.

Some flywheels and clutches may be balanced independently on special balance arbors (see Fig. 12-13). Remember, flywheels and harmonic balancers that are externally weighted as a part of crankshaft balance must be balanced as a part of the crankshaft assembly.

*FIGURE 12-13 Balancing a flywheel or clutch without a crankshaft in a Stewart-Warner balancer*

## SUGGESTIONS FOR MINIMUM BALANCING

Full engine balancing is an added expense paid for by the customer. While some can be readily sold on the benefits, others cannot. It is suggested that at least certain minimum corrections or checks be made on engine balance to ensure that engine rebuilding does not create unacceptable vibration.

First compare the relative weights of pistons in a set. Some manufacturers include specified weights in service manuals. It is recommended that a spread of five grams in a set be considered tolerable. Correction to this level means that perhaps only one or two pistons out of a set need weight removed.

On V-block engines, compare weights of replacement pistons to the original pistons or to manufacturers' specified weights. Remember that piston weights are included in crankshaft design and balance. Changing the weights of pistons appreciably will necessitate dynamic crankshaft balancing if the engine is to run smoothly.

Connecting rod weights should also be checked. This step is especially important if all connecting rods are not from the same original set. It is suggested that total weights be held within a seven-gram spread. In some cases the manufacturers provide guidelines for acceptable connecting rod total weights.

Clutches are another major source of engine vibration. An engine may run smoothly until a clutch is replaced and then develop a vibration. First, buy only the highest quality clutch pressure plates with a guaranteed close level of balance. Another option is to send out the clutch pressure plate to a machine shop equipped for balancing.

Following the recommendations given requires minimum tooling. A scale capable of weighing in grams is all that is required. While the engine will not be fully balanced, excessive variations in weights will have been detected and corrected prior to engine assembly.

# 13

# Engine Assembly

Assembly is the critical step in engine service. All of the painstaking attention to the inspection and reconditioning of component parts is lost unless equally painstaking care is given to assembly. Failure to check an incorrect fit or tighten a nut or capscrew properly can lead to major engine damage.

It is common in many situations today to have machining sent out to automotive machine shops. In this circumstance, it is especially effective to have the mechanic responsible for the engine assembly recheck the fits and clearances of remachined parts. This is an important quality control consideration because even the best of machinists are going to produce unsatisfactory work occasionally.

## FINAL CLEANING AND DEBURRING

Engine parts are generally degreased and decarbonized prior to machining. However, machining alone produces contaminants in the form of metal chips and abrasives, which can be highly destructive if left in the engine. This is also the appropriate time for removing sharp edges or burrs created as a result of disassembly, machining, or handling.

Be sure to scrub valve guides thoroughly with a bore brush and cleaning solution. Also deburr the edges of the combustion chamber to remove sources of preignition. This is especially important because resurfacing produces burrs around combustion chamber edges.

It is also recommended that head bolt holes be chamfered after resurfacing. This prevents the sharp edges of the threads from being pulled up into the cylinder head gasket when the head bolts are tightened.

Carbon, dirt, and other contaminants in threaded holes make correct torque readings virtually impossible during assembly. For this reason, it is also recommended that a tap or bore brush be run through head bolt and main bolt holes. A tap should not remove any significant amount of metal but should only scrape the threads clean.

Valve lifter bores and cam bearing bores should be inspected for burrs, corrosion, or varnish buildup, which might cause the incorrect fit of replacement parts. Valve lifters must rotate freely in their bores to maintain normal life for both the lifters and camshaft. Cam bearing bores should be clean and deburred so that new bearings can be pressed into place without shaving metal from the bearing shells or distorting them. These bores can be deburred by hand with emory cloth or in some cases by using a brake cylinder hone.

Once chamfering, tapping, and deburring steps are completed, the engine block should be thoroughly scrubbed in soap and water. Soap and water are more effective for loosening honing abrasives and cast iron particles than the usual cleaning solvents or solutions. Failure to clean the engine block thoroughly at this point will result in rapid piston ring wear and damage to engine bearings because abrasive contaminants will get into the engine oil.

## CYLINDER HEAD ASSEMBLY

As with the block, the cylinder head must be cleaned prior to assembly. It is especially important that valve guides be scrubbed clean. Remember that clearances are restored to new specifications and any abrasives remaining in a valve guide will cause valves to seize in the guides after only a few minutes of operation. The abrasives also combine with lubricants to form a lapping compound that causes severe wear in the valve guide.

It is recommended that an assembly lubricant be used on valve stems and in valve guides. SAE 90W gear oil will serve for this purpose if special products are not on hand. Also coat the valve face and seat. These steps prevent corrosion and galling (see Fig. 13-1) in a close-fitting valve guide. Lubrication is especially important if the cylinder head sits any length of time before use.

Installed height and stem length should have been corrected by this point, so care should be taken to keep valves in order. First, install a valve, spring shim, valve seal, spring, and retainer. Second, compress the valve spring and install the valve keepers. *Use care not to compress the spring any more than required,* or the valve seal will be crushed. Repeat this procedure until all valves are installed.

*FIGURE 13-1 A scored valve stem resulting from tight clearance, lack of oil, dirt, or a combination of these conditions*

## INSTALLING CORE PLUGS

Water leaks around core plugs (also called expansion plugs, soft plugs, freeze plugs, or Welsh plugs) are a common source of comebacks to the shop. Leaks may also lead to overheating and subsequent engine failure. Take the time to install them correctly the first time.

*FIGURE 13-2 This driver has a swivel end and drives against the outer edge of the core plug.*

*FIGURE 13-3 This driver has been selected to fit slightly loose against the inside of the core plug.*

First, measure the bore for each core plug. Order core plugs the same size as the bore. The core plugs are made to the bore size plus the correct amount for an interference fit. Deep-cup core plugs should be used whenever clearance and depth of the bore is adequate. Shallow core plugs are sometimes required. Stainless steel or brass core plugs are recommended for marine use.

Clean the bore with emory cloth to remove burrs and scale. It is recommended that the outer edges and the water jacket side of each core plug be coated with sealer to ensure sealing and to help prevent corrosion.

The core plugs should be driven into place until slightly below the surface of the block—generally even with the chamfered edge of the bore. Core plug drivers that drive against the outside edge of the plug are available (see Fig. 13-2). Other drivers should be selected so that they are approximately 1/32 inch (.8 mm) smaller than the inside diameter of the core plug (see Fig. 13-3). A driver that is too small in diameter will distort the core plug and the interference fit will be lost. A driver that fits too closely inside the core plug will be stuck in the core plug when it closes on installation.

## INSTALLING OIL GALLEY PLUGS

Some of the oil galleys are plugged with tapered pipe plugs. Tapered pipe threads lock and seal very securely with only moderate tightening. The only recommendation is to make sure threads are clean and coated with sealer on installation.

Core plugs are also used on oil galleys. One suggestion made earlier was to measure these plugs on removal so that replacements of the correct size may be ordered. If this step was omitted, measure the inside diameter of the bore. Order plugs for the measured size. Remember that the new plugs will come oversize the correct amount for the interference fit.

Select drivers in the same manner as for plugs in the water jackets. Drivers should drive against the outside shoulder of the plug or fit slightly loose (about 1/32-inch undersize) on the inside. Coat the outside edges of the plugs with sealer and drive them in slightly below the outside surface of the block.

Another precaution is to stake the plugs in place. This prevents high oil pressure from blowing the plugs out of their bores. This is done neatly with a cold chisel that has been blunted slightly across the cutting edge. The chisel is positioned across the opening to the oil galley and hit with a hammer, then turned 90° and hit again (see Fig. 13-4). This stakes the opening in four places (see Fig. 13-5).

*FIGURE 13-4 Using a blunt cold chisel to stake oil plugs in place*

*FIGURE 13-5 The appearance after staking*

## INSTALLING CAMSHAFT BEARINGS AND CAMSHAFT

The fitting of new cam bearings should be done carefully to ensure the proper fit of the camshaft through the bearings and the normal oiling of the engine. Cam bearings often vary in diameter according to the bearing position. They must be laid out carefully for position prior to installation. The bearings also frequently have oil holes that must be aligned with passages in the engine block so that oil can be pressure-fed through the camshaft bearings to other engine parts (see Fig. 13-6). Once the bearing positions and the alignment of oil passages are determined, the bearings are ready to be installed.

As mentioned, the bearing bores in the block should be cleaned and deburred. The outside of the bearing shells is coated with an assembly lubricant to make their installation easier and to prevent galling.

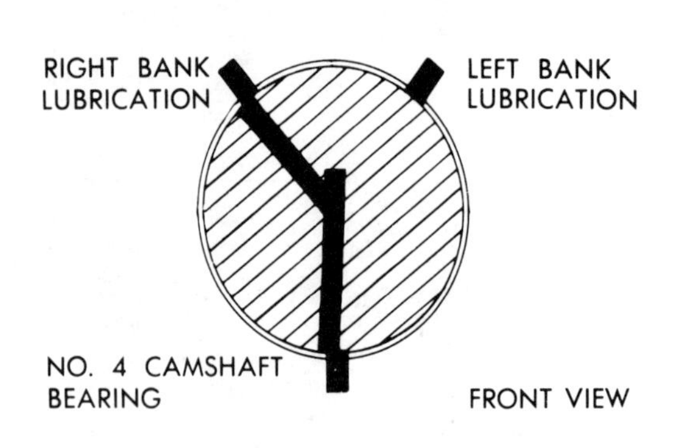


*FIGURE 13-6 Oiling through a cam bearing (Courtesy of Chrysler Corp.)*

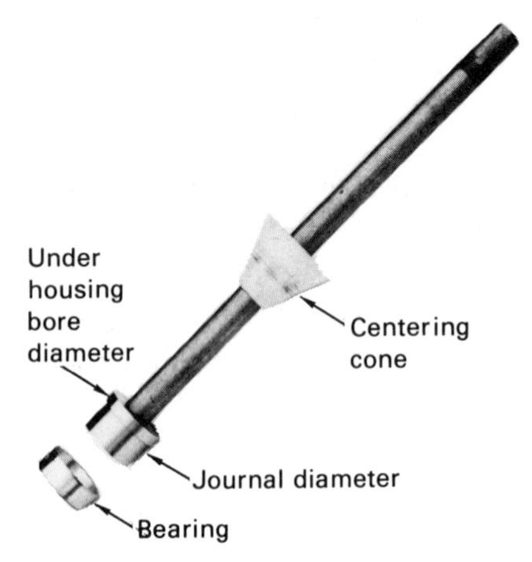


*FIGURE 13-7 A solid cam bearing driver. The smaller diameter is equal to the cam bearing journal diameter.*

Many production shops use solid cam bearing drivers made for each engine. Each driver has two diameters (see Fig. 13-7). The smaller diameter matches the camshaft journal diameter, and the larger diameter slips through the housing bore in the engine block. Typically, if the camshaft has only one journal diameter, only one driver is required for all bearings. If the camshaft journal diameters vary, one driver will be required for each bearing. The solid driver fits inside the bearing. An extension is used on the driver to reach all bearing positions (see Fig. 13-8). A centering cone may be used to help drive the bearing in straight. It is recommended that the edge of the bearing facing the driver be chamfered with a bearing scraper (see Fig. 13-9). This prevents bearing material from being "upset" (deformed to the inside) during installation and causing a binding on the camshaft.

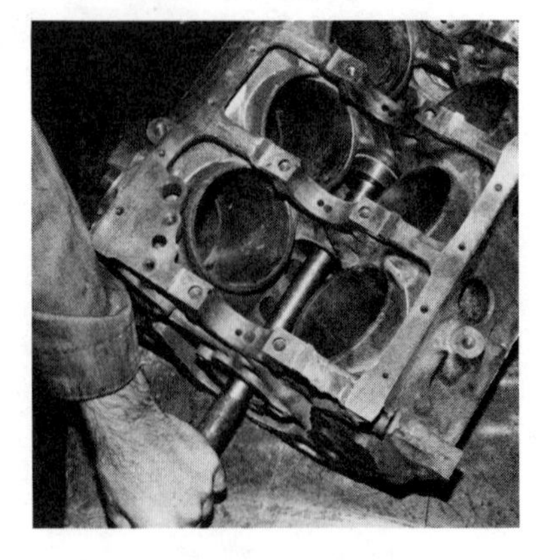

*FIGURE 13-8 Installing a cam bearing*

Be sure to check the alignment of oil holes through bearings after installation. This may be done by shining a light through oil passages. If the passage through the bearing is not accessible, a small inspection mirror may

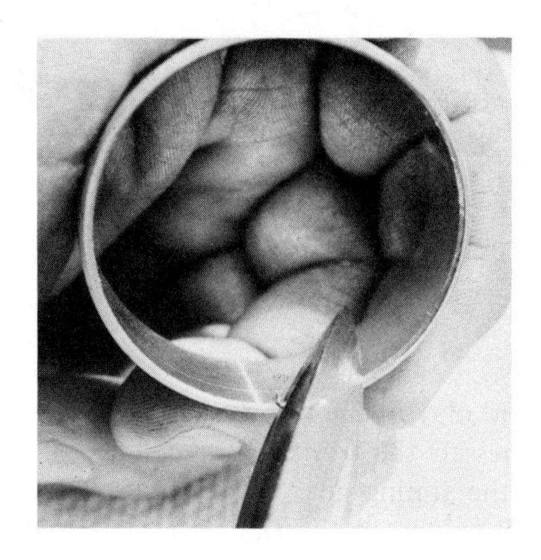

*FIGURE 13-9 Chamfering the edge on a cam bearing prior to installing it. This prevents the driver from upsetting the edge.*

be used to check for light. A bearing that is installed so that oil holes are blocked must be removed and discarded and a new bearing correctly installed in its place.

The next step is to check the fit of the camshaft in the new bearings. One easy way is to turn the block up on end. In this way, the camshaft may be lowered through the bearing bores with the least amount of interference between the cam lobes and the bearing surfaces. The fit may be checked with the block in a horizontal position, but care must be taken to prevent scoring the bearing surfaces with the cam as it is installed (see Fig. 13-10). The fit is considered acceptable if the camshaft can be rotated by hand. Of course, the cam journals are lubricated before installation. Any tight spots in bearings causing a binding of the camshaft may be removed by hand with a bearing scraper. If you suspect that the camshaft is not straight, check the alignment in V-blocks in the same manner as a crankshaft is checked (see Fig. 13-11).

*FIGURE 13-10 Installing a camshaft*

*FIGURE 13-11 Checking camshaft straightness in V-blocks*

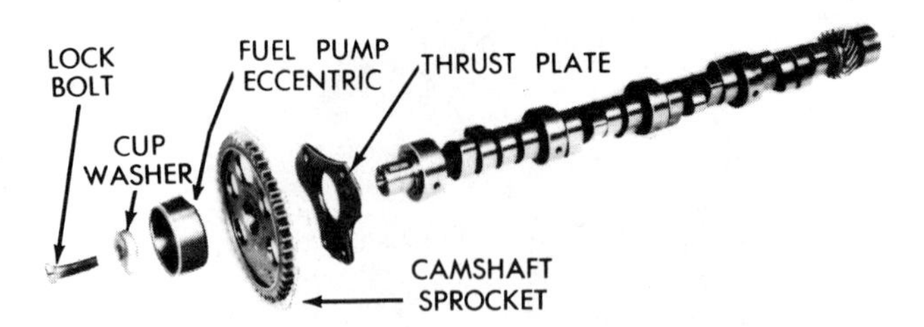


*FIGURE 13-12 A shoulder on the camshaft extends through the thrust plate to set end play (Courtesy of Chrysler Corp.)*

Some engines use a thrust plate to limit the end play of the camshaft. The thrust plate attaches to the front of the engine block behind the timing gear or sprocket. There may be a shoulder on the timing gear or sprocket, or on the camshaft, or there may be a spacer ring inside the thrust plate between the timing gear or sprocket and the front journal of the camshaft (see Figs. 13-12 and 13-13). The shoulder or spacer ring will be approximately .003-inch (.08 mm) thicker than the thrust plate to allow for that much end clearance.

A camshaft timing gear may be pressed off, and a replacement pressed back on. Be sure to align the thrust plate with the woodruff key during removal to prevent damage to the thrust plate. Both the thrust plate and timing gear must be aligned with the woodruff key for assembly.

Check camshaft end play with a feeler gauge between the thrust plate and the front

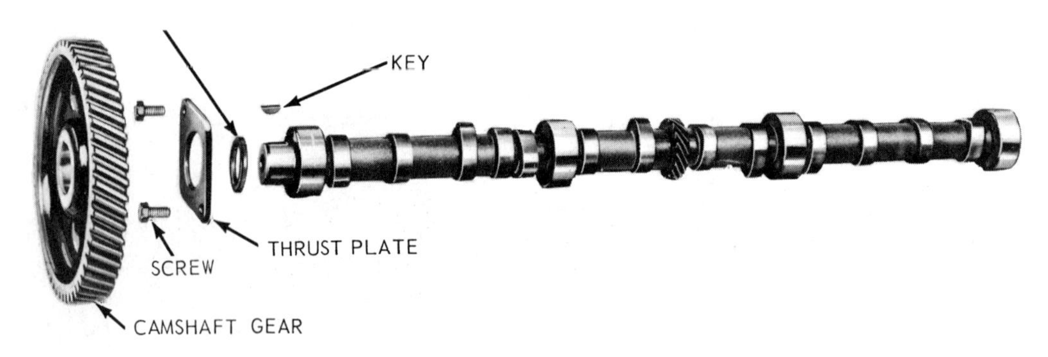


*FIGURE 13-13 A spacer ring and thrust plate used to control end play*

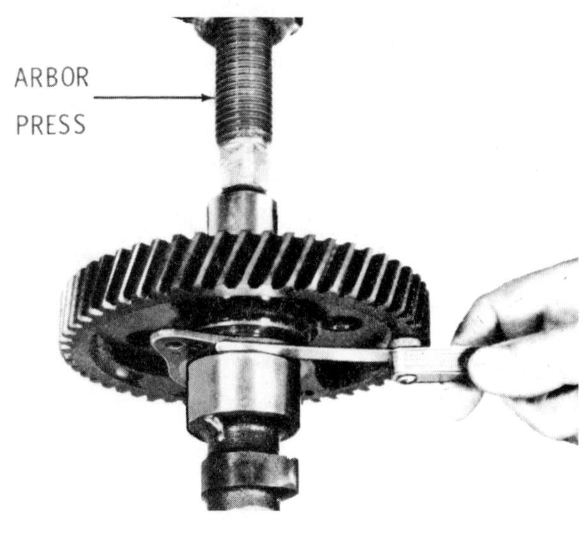


*FIGURE 13-14 Checking end play after pressing on a timing gear (Courtesy of Pontiac Motor Div.)*

journal of the camshaft (see Fig. 13-14). Excess end play is corrected by replacing the thrust plate.

The camshaft journals and cam lobes should be coated with an antiscuff lubricant for the final installation. The antiscuff lubricant protects the camshaft from extreme wear during the first minutes of engine operation. The rear cam plug should also be coated with sealer and installed at this time.

## FITTING THE REAR MAIN SEAL

Rear main seals of the rope type must be fitted correctly to prevent oil leaks or binding of the crankshaft. The procedure is begun by installing the rope packing in the groove and shaping it with a driver (see Fig. 13-15). The driver is the same diameter as the seal surface of the crankshaft. The ends are then trimmed flush with the engine block. Because it is easy to shape the seal in the block, this half of the seal can be removed from the block and installed in the rear main cap. The procedure can then be repeated for the block half of the seal. An alternate method of shaping the seal

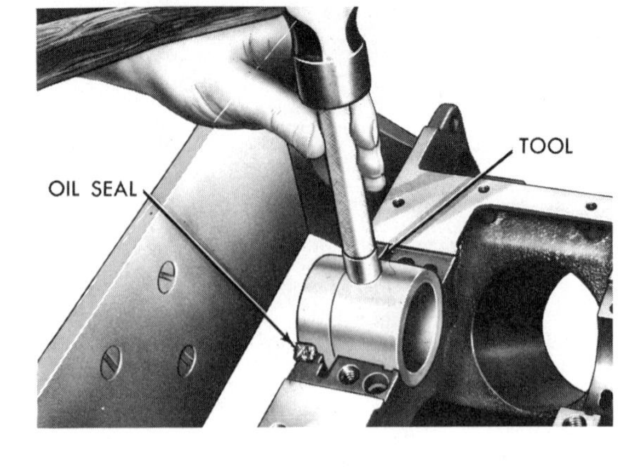


*FIGURE 13-15 Using a driver to shape the rear main oil seal (Courtesy of Chrysler Corp.)*

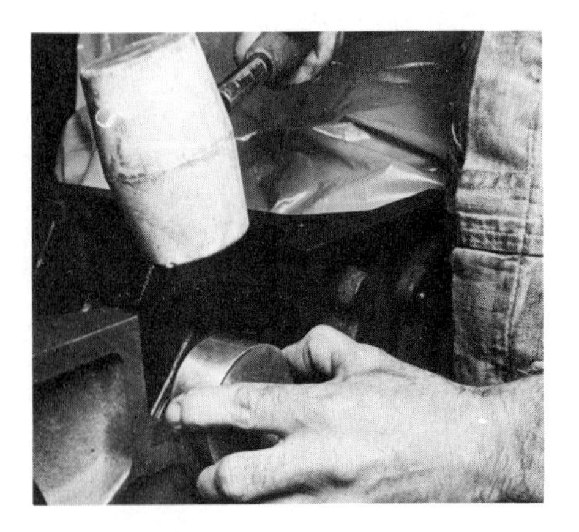

*FIGURE 13-16 Rolling a rear main seal into shape without round stock*

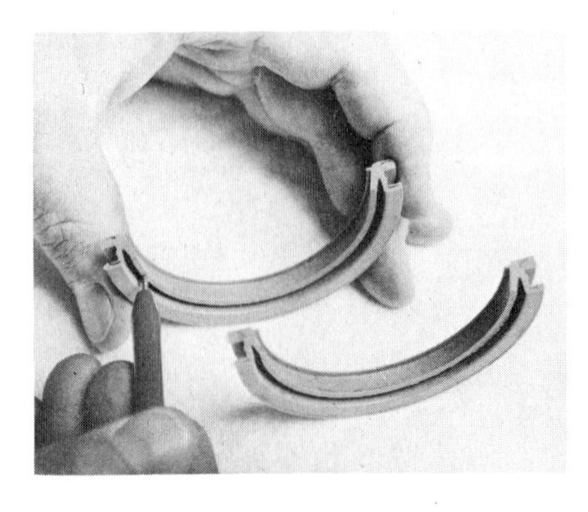

*FIGURE 13-17 A neoprene seal. The lip faces the engine oil.*

is to roll it into shape using a short length of round stock (see Fig. 13-16).

It is recommended that particular care be given to preventing the rear main seal from causing bearing misalignment. This is caused by rope material becoming jammed between the bearing cap and the engine block. Excess rope material can be detected by tightening the bearing cap in place, removing it, and checking for material between the cap and block surfaces. The excess material, if any, can be removed with a sharp knife or razor blade.

Of course, neoprene lip type seals are also very common today. Such seals present minimum problems during assembly. They need not be fitted because they are manufactured to fit both the block and the shaft. Be careful, however, to lubricate the seal with engine oil and to face the seal toward the oil inside the crank case (see Fig. 13-17).

Another source of leakage is the side seals of the bearing cap or seal cap (see Fig. 13-18). The side seals are often made of an absorbent material and are made to fit more closely by dipping them in oil immediately before installation. The oil causes them to swell. Another practice of some mechanics is to use a thin coating of gasket sealer across the rear edge of the bearing cap or seal cap to

*FIGURE 13-18 Installing side seals on a rear main cap*

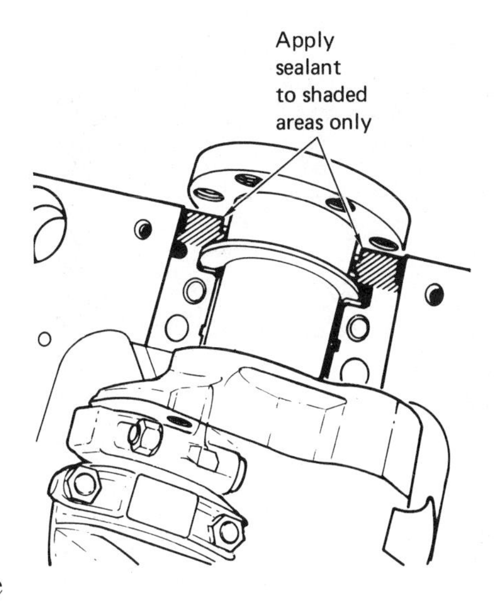


*FIGURE 13-19 Gasket sealer prevents oil seepage under the cap*

prevent oil seepage between the cap and the block (see Fig. 13-19).

## INSTALLING THE MAIN BEARINGS AND CRANKSHAFT

Main bearings should also be checked for location and position before installation. For example, main bearing sets have an upper half and a lower half. The upper half will have an oil hole, which must be aligned with oil passages in the block. The upper half may also have an oil groove while the lower half may not. Care must be taken to place the flanged main bearing (thrust bearing) in the correct location for the thrust surfaces of the crankshaft (see Fig. 13-20).

The bearings and the bore surfaces should be absolutely clean and dry. Contamination between the bearing shell and the surfaces of block or bearing cap will cause poor heat transfer and distortion of the bearing. The bearings are placed in position in the block and in the bearing caps and then lubricated. With the bearings in place and lubricated, the crankshaft is lowered carefully into place (see Fig. 13-21).

*FIGURE 13-20 The position of the thrust main bearing on one engine. This varies on different engines*

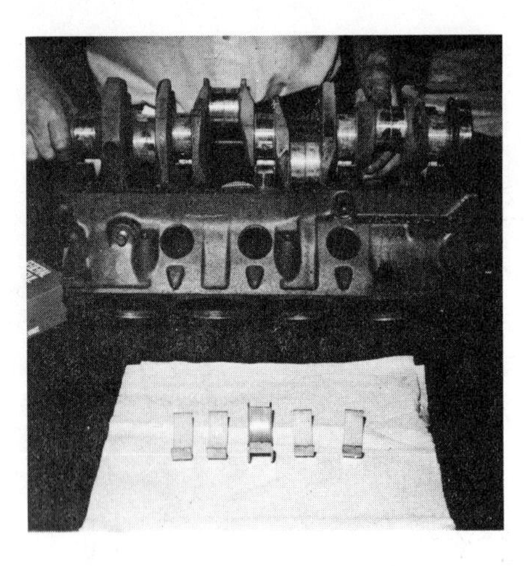

*FIGURE 13-21 Lowering a crankshaft into assembly*

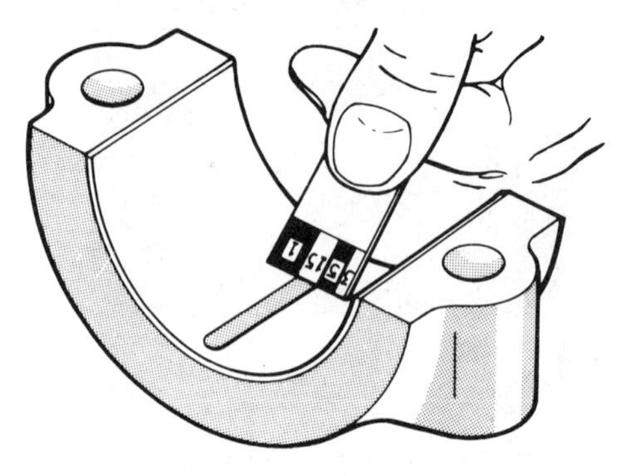


*FIGURE 13-22 Using Plastigage to check bearing clearance (Courtesy of Chrysler Corp.)*

Oil clearance may be checked at this point by using Plastigage. The plastic strip is placed across the crankshaft journal, and the bearing cap tightened to specifications. The bearing cap is then removed and the clearance checked against the graduated scale on the Plastigage package (see Fig. 13-22).

It is recommended that the main bearing capscrew threads be lubricated with engine oil before installation. Do not oil the internal threads because the hole may partially fill with oil, and hydraulic locking will prevent proper tightening of the capscrews. Tightening of the capscrews should also be done in stages of approximately 1/3 torque, 2/3 torque, and full torque. The crankshaft should be checked for free rotation after the bearing caps are tightened in place. Keep in mind that a properly fitted rear main seal of the rope type may cause a very slight binding of the crankshaft. A torque wrench can be used to check the turning force required. If a problem is suspected, recheck the rotation with the seal removed.

*FIGURE 13-23 Checking crankshaft end play with a dial indicator*

*FIGURE 13-24 Checking end play in an engine with a feeler gauge (Courtesy Buick Motor Division)*

Crankshaft end play is checked at this time. End play may be checked with a dial indicator and pry bar (see Fig. 13-23). The end play can also be checked with a feeler gauge between the thrust face of the crankshaft and the flanged thrust bearing (see Fig. 13-24).

A variation from these procedures is the installation of main bearings while the engine is in the chassis. This is referred to as *rolling* the bearings in or out. It is done by first backing off all main cap bolts a fraction of a turn so that the crankshaft can drop slightly. One main cap is then removed and a roll-out pin is placed in the crankshaft oil hole. The upper bearing is then removed by turning the crankshaft until the roll-out pin contacts the bearing and forces it to rotate with the crankshaft (see Fig. 13-25). A new bearing is installed by reversing the procedure. The lower bearing is placed in the cap and installed as usual. The process is repeated for each of the main bearings.

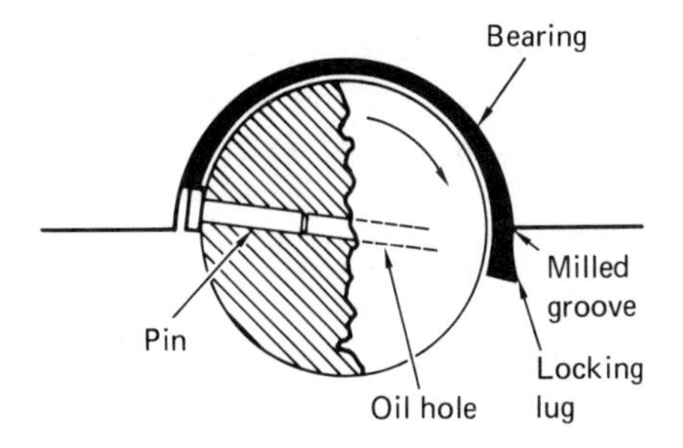


*FIGURE 13-25 Rolling out a main bearing in the car*

## SETTING VALVE TIMING

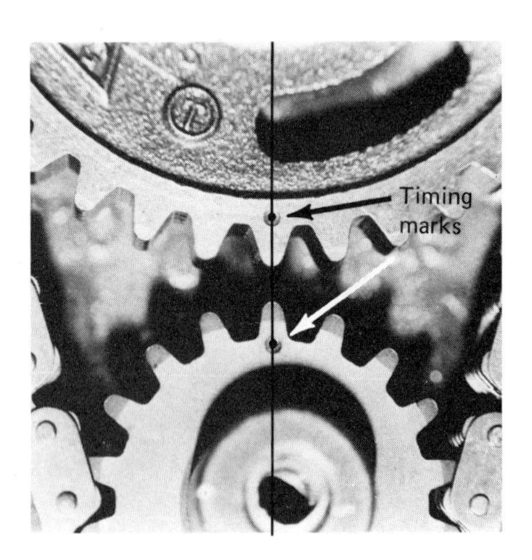


*FIGURE 13-26 Typical timing marks on timing chain sprockets. This varies on different engines*

The cam drive sprockets and timing chain must be installed in a specific position to set the engine valve timing. The crankshaft sprocket fits the crankshaft in only one position because of a key. The camshaft sprocket is twice the diameter of the crankshaft sprocket and is fixed in position on the camshaft by a key or possibly by a pin. There are timing marks on each of the sprockets. The valve timing is correct in many applications when the timing marks are on the centerline between the crankshaft and camshaft (see Fig. 13-26). However, check service manuals to be sure of valve timing marks because they do vary. Final assembly is done by sliding the two sprockets and the timing chain into position together. This minimizes the twisting or binding of the timing chain during installation.

In the case of timing gears, timing marks on the gears are typically aligned with the centerline between the camshaft and the crankshaft (see Fig. 13-27). As with timing chains and sprockets, be sure to check service manuals for the correct alignment of timing marks. A visual check of the cam lobe positions for number one cylinder should verify that both valves would be closed.

*FIGURE 13-27 Typical timing marks on timing gears (Courtesy of Pontiac Motor Div.)*

## INSTALLING PISTON RINGS

After the cylinder wall is well oiled, each piston ring is placed squarely in the cylinder, and the end gap is checked with a feeler gauge (see Fig. 13-28). The end gap is checked with the piston ring in the lower (unworn) area of worn cylinders. End gap may be checked anywhere in new cylinders. Common practice is to check for minimum end gap only—approximately .003 inch (.08 mm) per inch of cylin-

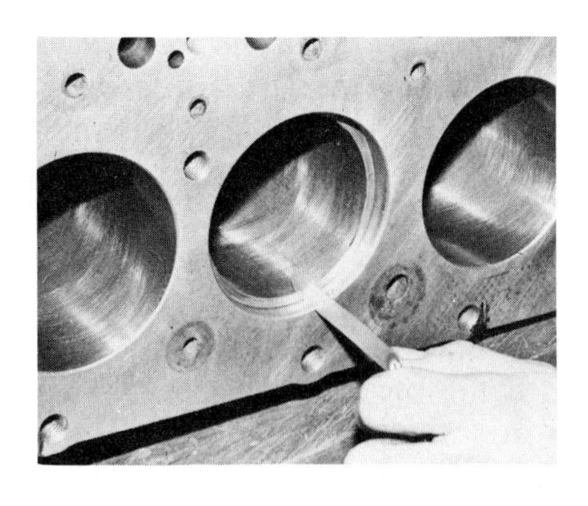

*FIGURE 13-28 Checking piston ring end gap*

der diameter. Without the minimum end gap, the piston rings will expand with heat and butt together, causing broken piston rings and scored cylinders. End gaps may be increased by filing (see Fig. 13-29).

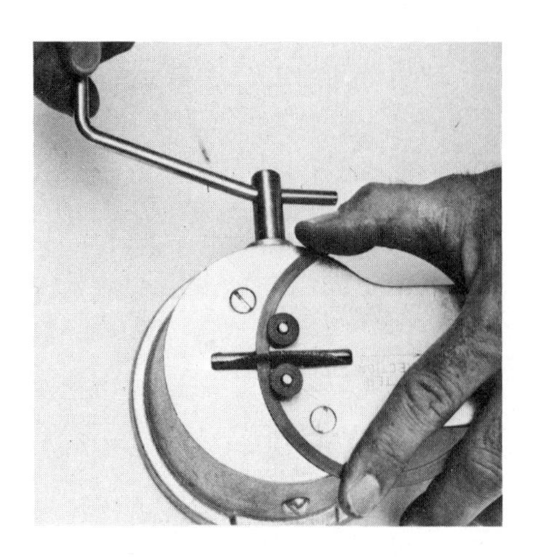

*FIGURE 13-29 Filing to increase piston ring end gap*

As mentioned, usually only minimum end gap is checked. Be aware, however, that some manufacturers specify a maximum end gap. It is recommended here that maximum end gap be checked to prevent rings for a smaller bore diameter from being installed. End gap increases about .003 inch (.08 mm) for each .001 inch (.03 mm) that bore size increases. This means that an incorrect ring set will have a dramatically increased ring gap. Rings that are run in oversize cylinders will not seal. Evidence of this, in the form of carbon, will be visible near the end gaps after the engine is run (see Fig. 13-30).

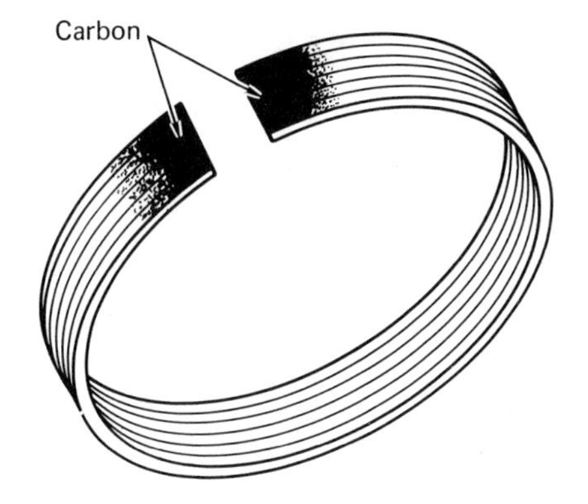


*FIGURE 13-30 Poor end sealing caused by the use of incorrect diameter piston rings*

*It is an absolute must to read piston ring installation instructions before installing the rings on the piston.* Installing a piston ring upside down or in the wrong ring groove can easily result in high oil consumption. Some examples of correct compression ring installation are shown in Fig. 13-31.

A typical ring installation procedure begins with the oil control rings. The ring expander is installed with the ends over the piston pin hole. It is critical that the ends of the expander not be trimmed or permitted to overlap because either condition will reduce oil control ring tension and cause oil consumption. The lower steel rail is then spiraled into place with the end gap approximately 2 inches to the left of the expander ends. The upper steel rail is spiraled into place with the end gap 2 inches to the right of the expander ends. This procedure is recommended as one method of ensuring that all the end gaps are offset.

It is also suggested that compression rings be installed with an expanding tool that ex-

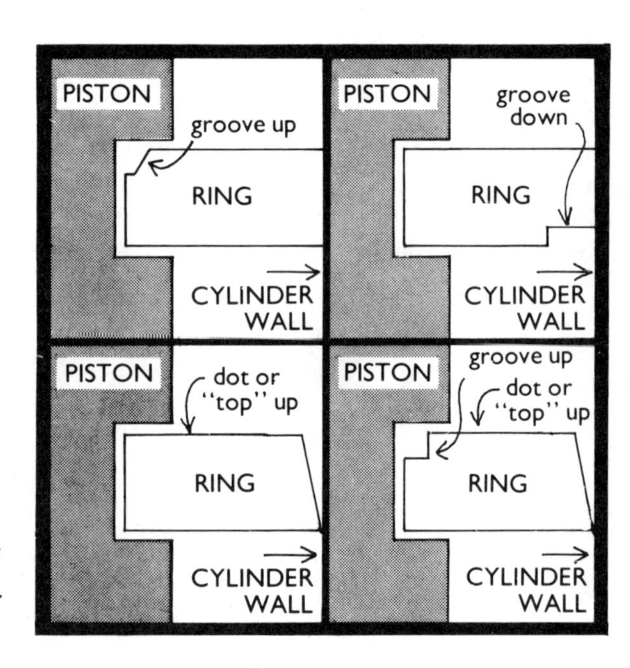


*FIGURE 13-31 Examples of correct piston ring installation. Always read instructions.*

pands the ring the minimum amount required to slip over the piston crown (see Fig. 13-32). These tools prevent breaking the rings on installation. The lower compression ring is installed first and the top compression ring last. Once installed, the piston rings are rotated so that all end gaps are offset. The piston rings and ring grooves should also be well oiled, possibly by dipping the piston in oil, and checking to see that the rings slide freely in their grooves.

*FIGURE 13-32 Using a ring expanding tool for installing rings on pistons*

## INSTALLING PISTON AND CONNECTING ROD ASSEMBLIES

Piston and connecting rod assemblies are first laid out in order according to cylinder number and the direction of installation. As mentioned before, the piston usually has a notch or other indicator that must point toward the front of the engine. Oil spurt holes or other references on connecting rods should also be checked for their proper direction.

Each half of each connecting rod bearing is now snapped into place, and the bearing surfaces lightly oiled. The cylinder is oiled and the crankshaft throw rotated to bottom dead center. Rubber tubing is placed over each rod bolt, and a ring compressor is clamped over the piston rings. The piston and connecting rod are now pushed or butted lightly into assembly with a soft-faced hammer or a hammer handle (see Fig. 13-33). The connecting rod should be guided carefully over the crankshaft throw to prevent nicking the surface. The rubber tubing may then be removed, and the rod cap and bearing tightened into place. Next the rod bearing oil clearance is checked with Plastigage as were the main bearings. It also is recommended that new self-locking

*FIGURE 13-33 Using a ring compressor to install piston and rod assemblies*

*FIGURE 13-34 Checking rod side clearance*

connecting rod nuts be used to prevent backing off caused by engine vibration. If new lock nuts are unavailable, an anerobic adhesive (such as Loc-Tite) may be used on the threads.

Once the connecting rods are tightened into position, the side clearance between connecting rods must be checked with a feeler gauge (see Fig. 13-34). Frequently, only the low limit of side clearance is checked. It should be noted that bent or twisted connecting rods may well cause side clearance to be tight in spots, so be sure to check for clearance in two or three locations. If side clearance is found to be tight, the alignment should be rechecked. If necessary, side clearance may be increased by lightly sanding or filing the side faces of the connecting rods.

Excessive side clearance causes increased oil throw-off onto cylinder walls and can cause increased oil consumption. However, this condition is sometimes not corrected because reducing side clearance requires replacement of at least some connecting rods.

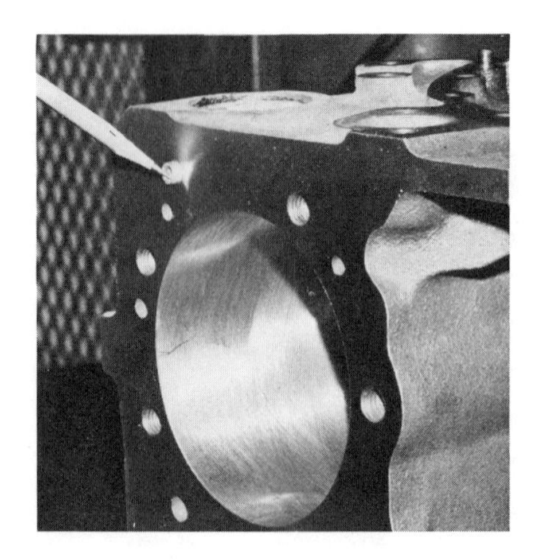

*FIGURE 13-35 Typical locating pins on an engine block to position cylinder heads and gaskets*

## ASSEMBLING CYLINDER HEADS TO ENGINE BLOCKS

Many checks and procedures have been taken prior to this point to ensure positive gasket sealing. Surfaces are flat and clean, and head bolt threads are clean. The sharp edges around combustion chambers have also been deburred. Be sure that locating pins have been replaced in the engine block (see Fig. 13-35). These keep the cylinder head and gasket in alignment during assembly.

Now check the cylinder head gasket for correct position for assembly. Some gaskets are marked with TOP or FRONT to indicate position. On some engine blocks, the locating pins are offset to different positions at each end of the block so that gaskets will fit in only one position. Keep in mind that coolant circulation is metered by the passages through head gaskets and improper assembly could cause an engine to overheat. Oil passages to rocker arms may also go through head gaskets and improper assembly may block these passages.

Head bolts should be checked for length, position, and thread condition. Some head bolts may have special configurations such as studs at the top end in order to mount accessories. Lubricate the threads of the head bolts with engine oil and place them in position. *Do not oil internal threads;* oiling may cause hydraulic locking when the head bolts are tightened.

Antiseize compounds are sometimes used on head bolts in aluminum blocks. Consider that these compounds reduce friction at the surfaces of mating threads. In fact, friction is reduced to such an extent that the expected clamping force is obtained at about two thirds of specified torque. This means that specified torque should be reduced approximately thirty percent if antiseize compounds are used. Failure to reduce torque will cause overtightening and broken or stripped head bolts and internal threads.

Head bolts are tightened in two or possibly three stages. For example, they may be tightened to 1/3 of specified torque, 2/3 torque, and then full specified torque. A pattern must also be followed for the sequence of tightening head bolts. Usually the pattern begins near the center of the head and works alternately left and right around each combustion chamber. Be sure to check service manuals for the exact procedure on each engine.

The valve lifters and push rods are now installed. Coat the bases of valve lifters and tips of push rods with antiscuff lubricant.

## INSTALLING ROCKER ARMS

When preparing to assemble rocker arms and shafts or to attach rocker arm assemblies to the engine be sure to check rocker arm positioning over valve stems. Many engines use rocker arms that are offset to the right or left (see Fig. 13-36). Incorrect assembly of the rocker arms will cause the faces of the rocker arms to misalign with valve stems.

Check the positioning of rocker arm shafts as well. Most of the shafts must position in a certain way to ensure adequate lubrication. Follow assembly instructions for the particular engine so that rocker arm oil holes are correctly positioned (see Fig. 13-37). Make sure that oil plugs are installed in the ends of shafts and that all parts are well lubricated prior to assembly.

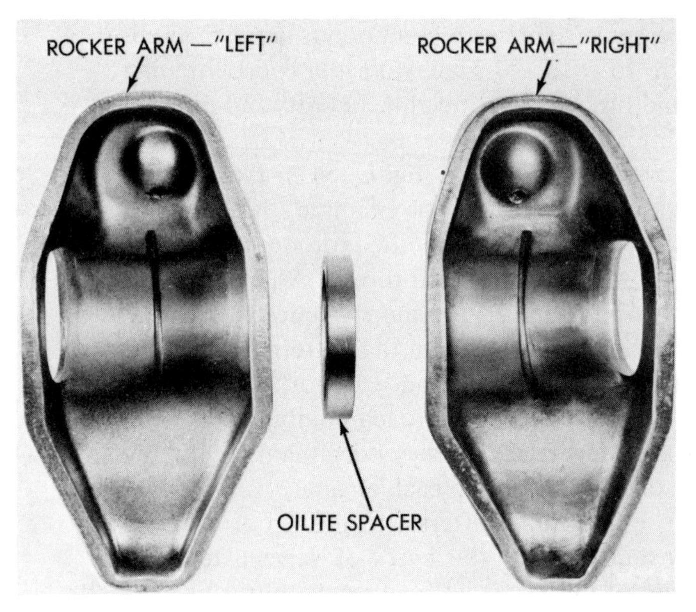


*FIGURE 13-36 Offset position of rocker arms (Courtesy of Chrysler Corp.)*

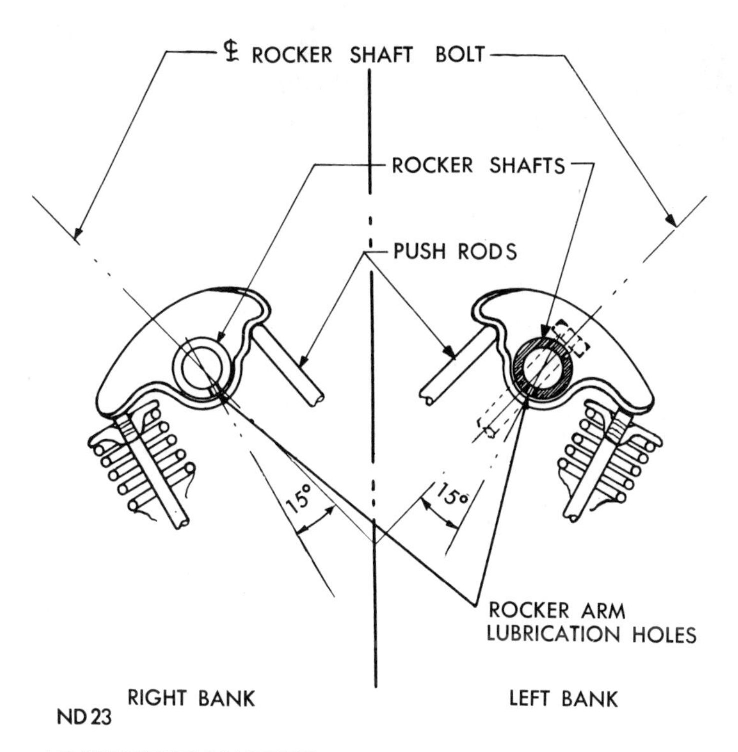


*FIGURE 13-37 Typical instructions for correct assembly of rocker arm shafts (Courtesy of Chrysler Corp.)*

## ADJUSTING VALVES

The valve lash may now be adjusted to clearance specifications on solid valve lifter engines (see Fig. 13-38). One method is to turn one cylinder at a time up to top dead center (TDC) on the compression stroke and adjust both valves. While this procedure is relatively easy to follow during engine assembly, there are faster methods given in service manuals for each particular engine. The faster methods call for adjusting valves on different cylinders at each crankshaft position. The procedures vary according to crankshaft design.

Engines with hydraulic valve lifters and adjustable rocker arms are adjusted by two basic methods. One is to adjust each rocker arm to zero lash while the valve is closed. Final adjustment is made after running the engine by backing off each rocker arm until it rattles, tightening it until quiet, and then tightening it

*FIGURE 13-38 Adjusting valve clearance*

*FIGURE 13-39 Collapsing a hydraulic valve lifter to check rocker arm adjustment*

1/2 to 3/4 of a turn more. Another method is to collapse each hydraulic valve lifter by prying down on the push rod end of the rocker arm and then checking for a specified clearance between the valve stem tip and rocker arm face (see Fig. 13-39). Check service manuals to see which method applies.

Obviously, nonadjustable rocker arms are tightened down only. However, remember that valve stem length is critical because it is possible for valves with long valve stems to be held open by the rocker arms.

## INSTALLING THE OIL PUMP

The engine assembly is nearly completed at this point. One of the last steps is preparing for the installation of the oil pump. Some mechanics check the pump operation and prime it by dipping the pump pickup in oil and turning the pump by hand. Other procedures call for packing the pump in petroleum jelly to ensure that it will be primed when the engine is started. Check service manuals for specific recommendations.

There are other perhaps less obvious checks that must also be made. First, the oil pump pickup should be within 1/4 inch (6 mm) of the bottom of the oil pan. Once the position of the pickup is checked, the pickup tube should be securely fastened to the pump body. In some cases, the pickup tube bolts to the pump, and no particular problems are involved. In other cases, the pickup tube presses into the pump body. As mentioned earlier, it may be desirable to braze a press-fit pickup tube into the pump body to prevent it from leaking or slipping out of assembly. Just be sure to remove all internal parts before brazing. Once the position and attachment of the pickup tube is checked, the pump is ready for installation in the short block. If the oil pump pickup should fall out, oil pressure will be lost. If the pickup should fit loosely, it will draw in air, and engine lubrication will suffer because of aerated oil. If it is too high, oil will wash away from the pickup on braking, acceleration, and cornering.

It should also be noted that some engines use an intermediate shaft to connect the oil pump to the distributor (see Fig. 13-40). Some shafts, and retaining clips or sleeves, must be installed with the oil pump from the lower side of the engine. Failure to catch this detail may mean having to remove the oil pan from the assembled engine.

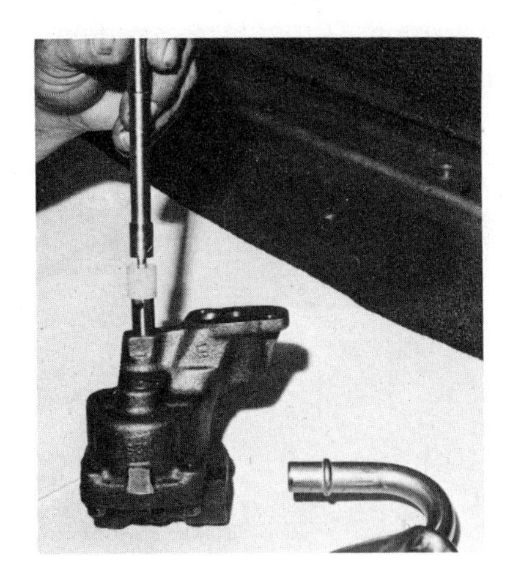

*FIGURE 13-40 A typical oil pump drive system using a sleeve around the drive*

## PREOILING THE ENGINE

Preoiling internal parts prevents scuffing bearings, cylinders, and camshafts. A common cause of damage, especially to bearings, is scuffing that occurs during dry starting of new or overhauled engines.

The preoiling may be done using the engine's own oil pump. In this case the oil pan is installed and filled with oil. Valve lifters must also be installed to block oil passages through lifter bores. The oil pump is driven by hand with a speed handle (see Fig. 13-41). Using a drill motor is not recommended because the pump may be damaged internally. Hand rotation will provide 20 psi or more of oil pressure. The direction of rotation must be the same as for the distributor. Rotate the engine to several positions to be sure oil circulates to all engine parts. Of course, this procedure also primes the oil pump so that the engine will have immediate oil pressure when started.

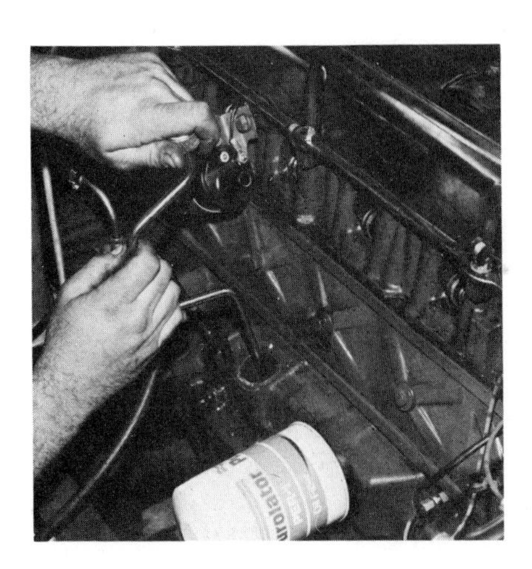

*FIGURE 13-41 Priming an oil pump and pre-oiling an engine by hand*

A prelubricator may also be used. The prelubricator is a pressurized oil tank connected to an engine oil galley by a line (see Fig. 13-42). Usually a connection can be made

*FIGURE 13-42 A pre-lubricator connected to an engine*

through the location for the oil pressure switch. While this procedure oils the engine parts, it may not prime the oil pump. It is suggested that the prelubricator be left connected while the engine is cranked over by the starter motor in order to prime the pump. The prelubricator may then be disconnected, the oil galley plugged, and the engine started.

A prelubricator is also an excellent diagnostic tool. The oil pan can be removed and excess oil flow through bearings or leaking oil galleys can be seen visually (see Fig. 13-43). Oil circulation through various passages can also be checked.

Take careful note that hydraulic valve lifters were not filled with oil prior to assembly. While it is common practice to fill lifters first, it has been found that overfull valve lifters may hold valves open and even cause bent valves if they should interfere with the pistons. Rather than attempt to determine which engines present the greater risks, it is recommended that hydraulic valve lifter oiling be left until the priming and preoiling steps. This works well because push rods, rocker arms, and valve springs limit plunger travel in the lifters and prevent overfilling.

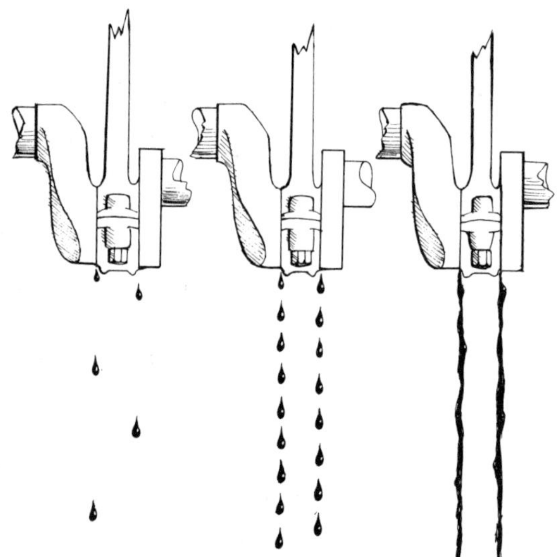

*FIGURE 13-43 Checking for differences in oil flow through engine bearings with a prelubricator*

About the only alternative to preoiling an engine is packing the oil pump with petroleum jelly. Packing the pump primes it and at least assures that it will begin working immediately when the engine is first cranked over. Of course, all other engine parts must be well lubricated on assembly. This procedure is desirable when the oil pump is driven directly from the camshaft or crankshaft (because the pump cannot be turned without cranking the engine) and there is no prelubricator on hand.

## HINTS ON GASKETS AND SEALS

Several points have been made already regarding cylinder head gaskets. As mentioned, they must be installed in the correct position in order not to restrict water or oil passages. Another suggestion is to apply gasket sealer to the gasket and not to the engine block or cylinder heads. This keeps sealer out of cylinders or combustion chambers. Spray cans make the neat application of sealer easy. Just hang up the gasket and spray it before assembly.

It is recommended that silicone sealer be applied to any intersections of gaskets such as at the corners of oil pan gaskets. Silicone sealer works very well in those locations because the sealer readily flows into gaps where gaskets come together. Also watch the corners of the intake manifold where gaskets, the engine block, and cylinder heads all come together. It is recommended that sealer be applied at points where these parts intersect (see Fig. 13-44).

Silicone sealers also work especially well around water passages. For example, it is good practice to apply a bead of sealer around the water passages on intake manifold gaskets (see Fig. 13-45).

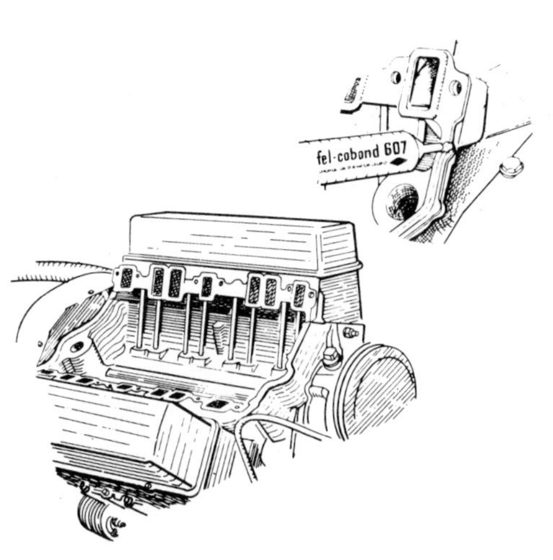


*FIGURE 13-44 Using sealer at the corners of the intake manifold gaskets (Courtesy of Fel-Pro Inc.)*

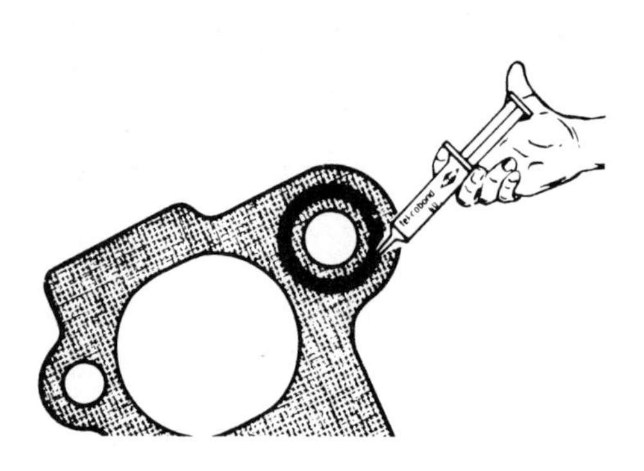

*FIGURE 13-45 Using sealer around water passages through intake manifold gaskets (Courtesy of Fel-Pro Inc.)*

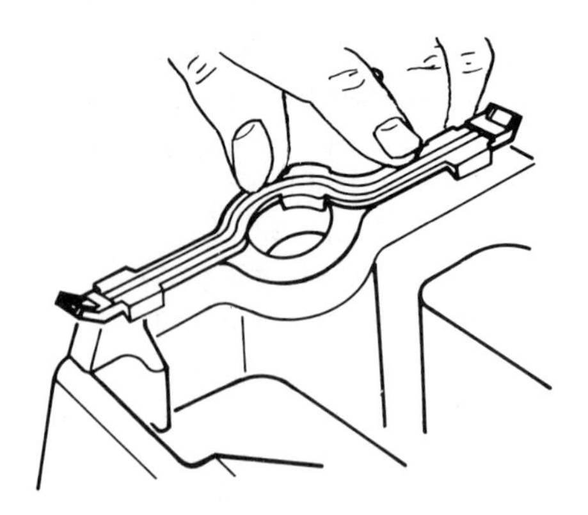

*FIGURE 13-46 A close-up view of an intake manifold end gasket (Courtesy of Fel-Pro Inc.)*

The end gaskets under intake manifolds sometimes cause difficulty during assembly (see Fig. 13-46). The problem is that the gasket will sometimes slip out of position when the manifold is tightened. This may be prevented by using a contact cement sealer. Apply sealer to the gasket and to the block surface under the gasket. Allow the sealer to dry in the air for a few minutes before assembly. When the gasket is installed, it will be cemented in place.

All sheet metal should be straightened before assembly. Watch especially for distortion around the bolt holes in oil pans and valve covers from overtightening. Check the fit of gaskets and alignment with bolt holes. Remember that cork gaskets will shrink because of drying. They can be made to expand by wetting them with warm water. Be careful not to overtighten bolts on assembly. Just snug them lightly to approximately 100 inch-pounds during assembly and retighten them after the engine warms up.

Timing cover seals are another source of oil leaks. Be sure to apply gasket sealer around the seal case before installing it in the timing cover. Also lubricate the seal to prevent dam-

*FIGURE 13-47 A worn seal surface on a harmonic balancer*

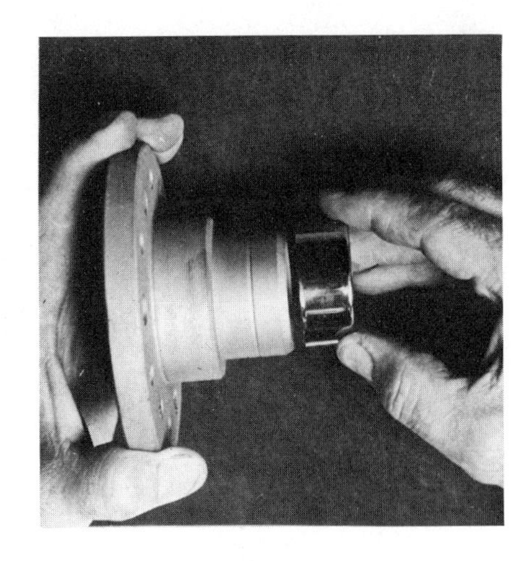

*FIGURE 13-48 A sleeve used to repair the seal surface*

age during the first few minutes of operation. A deep groove sometimes develops on the vibration damper seal surface, making an effective seal at the timing cover very difficult (see Fig. 13-47). Check to see whether the replacement seal locates in a wear groove. If it does, one trick is to reposition the seal to a slightly different depth on installation. For some applications, a thin wall sleeve is available to slip over the vibration damper (see Fig. 13-48). The vibration damper may have to be replaced if these methods will not work.

# 14

# Engine Installation and Break-In

The actual "how-to" information for installing an engine varies with each make and model. Because of differences in vehicles, it is suggested that service manuals be checked for exact procedures. In addition, it is suggested that some sort of check list be used to make a final inspection of an engine installation. Such an inspection can ensure normal and safe operation of the reconditioned engine. Engine break-in and follow-up services are also important to long-term satisfaction with engine service.

## ENGINE INSTALLATION CHECK LIST

The items on such a check list may be inspected during disassembly, engine service, or after installation. However, all items must be checked prior to returning the vehicle to the owner.

1. Engine mounts for tightness or breaks
2. Exhaust leaks
3. Fuel leaks
4. Radiator hose condition
5. Heater hose condition

6. Hose clamp tightness
7. Condition and tightness of drive belts
8. Transmission linkage adjustment
9. Transmission oil cooler line leaks
10. Circulation of coolant through radiator
11. Pressure cap operation
12. Radiator and heater core leaks
13. Oil leaks
14. Oil pressure
15. Cylinder compression
16. Condition of points, distributor cap, rotor, secondary cables, and spark plugs
17. Adjustment of dwell, timing, and carburetor
18. Vacuum leaks
19. Battery charge and cable connections
20. The proper routing of all hoses and wiring
21. Cleanliness of the engine compartment and the inside and outside of the vehicle

## ENGINE BREAK-IN

For years it was common practice to break in engines on lightweight, nondetergent oil. Currently it is recommended that the crankcase be filled with an oil having the service rating intended for the engine. This recommendation is based on the fact that many emission controlled engines develop tremendous heat, and oil with a lesser service rating than specified will break down and will not maintain normal oil pressure and lubrication. This oil breakdown can lead to early failure in a new engine.

In addition, oils with lesser service ratings do not provide adequate scuff protection for the camshaft.

The engine should be run at a high idle speed for the first few minutes of operation to ensure adequate oil throw-off onto cylinders and oil circulation to all moving parts. The vehicle should then be run on the road for the initial seating of the piston rings. This is done by cycling the engine speed—first accelerate, then decelerate until oil smoke is no longer apparent at the exhaust pipe. The cycling of the engine causes rapid ring seating. Gas pressure forces compression rings against the cylinder walls on acceleration. The increased forces cause seating to take place more rapidly. On deceleration, oil is drawn up onto the cylinder walls because of the low pressure (vacuum) above the piston. This prevents cylinder scuffing or scoring during cycling. The engine idle speed may be set back to specifications after the engine is cycled.

Driving at high speeds may overheat new bearings. New bearings are relatively soft and deform easily. This softness is necessary to permit bearings to conform to the shape of housing bores and rotating shafts. The bearings conform to shape and harden after the first several hours (500 miles) of engine operation and are not as easily damaged. Long periods at idle should be avoided because oil throw-off onto cylinder walls is reduced. Overheating may also occur during long periods at idle.

It is generally recommended that the owner drive the car "normally" during the first 500 miles. High speeds and long periods at idle are to be avoided. The owner must also be cautioned to watch engine temperature indicators because new engines tend to run hotter because of increased friction and efficiency.

## FOLLOW-UP SERVICES

The owner should be cautioned about the importance of follow-up service. The oil and oil filter should be changed after break-in because abrasives and metallic particles picked up during engine service work free during the break-in period. Cylinder heads and intake manifolds should also be retorqued after break-in to compensate for changes in gaskets caused by heating and cooling cycles.

It will be found that valves will recess slightly into valve seats during break-in. Because of this recession, valve adjustment should be rechecked in all engines without hydraulic valve lifters. Readjusting valves with hydraulic valve lifters will not be necessary because plunger travel will allow for valve recession.

Another recommendation is to review another check list to catch problems that might cause failure or a comeback. Of course, careful attention should be given to customer complaints or comments regarding the operation of the car. Such a check list might include the following:

1. Customer complaints
2. Smooth operation and normal power
3. Engine or exhaust noises
4. Oil pressure and water temperature indicators
5. Automatic transmission operation and shift points
6. Manual transmission and clutch operation
7. Fuel, oil, and coolant leaks

## *REVIEW QUESTIONS*

**Directions:** Mark the letter choice of the best answer to each question.

*Chapter 1, Engine Theory*

1. A four-cylinder engine fires one cylinder every
   a. 90°
   b. 120°
   c. 180°
   d. 360°

2. The intake valve opens
   a. at TDC
   b. just after TDC
   c. just before TDC
   d. just after BDC

3. On the exhaust stroke
   a. both valves are open at TDC
   b. the exhaust valve is closed at TDC
   c. the intake valve is open at BDC
   d. both valves are closed at BDC

4. Duration is the time that
   a. a valve is closed
   b. both valves are closed
   c. both valves are open
   d. a valve is open

5. Cam lift is equal to the
   a. radius of the base circle
   b. diameter of the base circle
   c. maximum measurement of the cam lobe minus the radius of the base circle
   d. maximum measurement of the cam lobe minus the diameter of the base circle

6. Valve lift on OHV engines
   a. exceeds cam lift
   b. is equal to cam lift
   c. is less than cam lift
   d. is equal to the radius of the base circle

7. Valve clearance is adjusted when valve lifters are on the ________ of the cam
   a. opening ramp
   b. closing ramp

c. base circle
d. nose

8. The check valve in a hydraulic lifter operates as follows:
 a. check valve closed when the engine valve is open
 b. check valve open when the engine valve is open
 c. check valve open when the engine valve is closed
 d. a and c above
 e. b and c above

9. Predetermined leakdown in a hydraulic valve lifter is
 a. due to excess plunger to body clearance
 b. due to defective check valve operation
 c. caused by dirt or varnish build-up
 d. normal

10. Oil baffles in the oil pan are used to
 a. keep oil in the sump on braking, acceleration, or cornering
 b. minimize air turbulence in the sump
 c. allow oil to return more quickly to the sump
 d. prevent overfilling the sump

11. An oil spurt hole in a connecting rod
 a. permits oil to flow into the rod bearing
 b. primarily lubricates the camshaft
 c. primarily lubricates the cylinders
 d. prevents excess oil pressure in the rod bearings

12. If the oil pump drive-shaft extension is left out of an engine, the engine would most likely
 a. start and run with oil pressure
 b. start and run without oil pressure
 c. start and run with spray oiling only
 d. not start

13. In a full-flow filtering system, a plugged oil filter causes oil to
 a. by-pass the filter and return to the sump
 b. by-pass the filter and lubricate the engine
 c. by-pass the filter and return to the inlet side of the oil pump
 d. stop flowing

14. The heaviest duty gasoline engine oil listed below has a service rating of
 a. MS
 b. CD
 c. SE
 d. SAE 20-50

*Chapter 1, Engine Theory* (continued)

15. A crankshaft bearing with .001″ clearance has an oil film ________ thick
    a. .0005″
    b. .001″
    c. .002″
    d. .004″

*Chapter 2, Engine Diagnosis*

1. Oil passing through valve guides may be due to
    a. valve seals damaged by heat
    b. excess valve stem to guide clearance
    c. manifold leaks
    d. a and b above
    e. b and c above

2. Oil burning under acceleration (or load) is caused by
    a. worn valve guides
    b. poor valve guide sealing
    c. poor valve sealing
    d. worn piston rings

3. Combustion gases in the cooling system are detected
    a. by drawing coolant through a "Bloc-check" solution
    b. by drawing vapors from the radiator through a "Block-check" solution
    c. only by an engine overheating condition
    d. only by coolant loss

4. A compression test can be used to determine
    a. which valve leaks, intake or exhaust
    b. if valve guides are leaking
    c. rings are worn
    d. all of the above
    e. none of the above

5. A cylinder leakage test can be used to determine if
    a. an exhaust valve leaks
    b. an intake valve leaks
    c. worn valve guides are leaking
    d. all of the above
    e. none of the above

6. At TDC on the exhaust stroke, the valves should be as follows
    a. exhaust closed, intake open

b. exhaust open, intake closed
c. exhaust closed, intake closed
d. exhaust open, intake open

7. Engine condition is being diagnosed. Timing chain backlash is measured by
   a. measuring chain slack with a rule
   b. turning the engine forward to TDC and backwards until the rotor moves and reading the backlash at the timing marks
   c. using a timing light to see if timing advances and retards on deceleration
   d. any of above
   e. none of above

8. A low and steady vacuum gauge reading is caused by
   a. poor ring sealing
   b. a vacuum leak
   c. retarded ignition timing
   d. any of above
   e. none of above

9. An engine knock is decreased by disconnecting a spark plug wire. The problem on that cylinder is
   a. excess main bearing clearance
   b. excess rod bearing clearance
   c. excess piston pin clearance
   d. a defective hydraulic valve lifter

10. An engine knock is diminished when the operating temperature reaches normal. The reason for the noise when cold is
    a. excess piston clearance
    b. excess rod bearing clearance
    c. excess piston pin clearance
    d. a hydraulic lifter which bled down

*Chapter 3, Engine Disassembly*

1. Valve spring retainers are freed from valve keepers by
   a. using penetrating oil
   b. striking the tip of valve stem with a hammer
   c. striking the spring retainer with a hammer
   d. driving straight down on the spring retainer using a piston pin or similar driver

2. Valves with mushroomed tips are best removed from the valve guide by
   a. filing or grinding away the mushroom
   b. driving out the valve guide with the valve

c. driving the valve through the valve guide
d. soaking the valve guide and valve stem with penetrating oil

3. Rods should be numbered
   a. as they are removed from the cylinder
   b. while clamped in a vise
   c. after fitting to replacement pistons
   d. before loosening from the crankshaft

4. When removing the ring ridge from an engine to be overhauled, removal should not extend into the area of
   a. piston travel
   b. ring travel
   c. cylinder honing
   d. oil control ring travel

5. To protect the crankshaft while removing or installing piston and rod assemblies, you should
   a. place bearings on crankpins
   b. place bearings in rod
   c. remove or replace rods with crank out of block
   d. place rubber tubing on rod bolts

6. Before removing a camshaft the
   a. valve lifters should be removed or moved out of the way by turning the cam one full turn with the engine upside down or on end
   b. cam thrust plates should be loosened
   c. timing chain should be removed or released from the camshaft sprocket
   d. all of the above
   e. none of the above

7. Before disassembly, main bearing caps should be
   a. marked with letters
   b. marked with arrows
   c. numbered from front to rear
   d. numbered from rear to front

8. Care should be taken in removing cam bearings not to
   a. damage the bearings
   b. damage the bearing housing bore in the block
   c. remove cam plugs from the block
   d. use cam bearing drivers when a punch or chisel will do

9. Threaded oil plugs are best removed by
   a. using an impact driver

b. using a breaker bar
c. heating around each plug before attempting to remove it
d. heating each plug before attempting to remove it

10. Core plugs in water jackets should be removed
    a. by heating them first
    b. by unscrewing them at room temperature
    c. by driving the plug inside and prying it back out
    d. only if leaking

*Chapter 4, Cleaning Engine Parts*

1. In order to adequately clean V-block intake manifolds
    a. carbon must be removed from cross over passages and from under shields
    b. they must be descaled
    c. carburetor mounting studs must be removed
    d. all of above
    e. none of above

2. Before reassembly, oil pump pick-ups must be
    a. washed in solvent
    b. hot tanked
    c. bead blasted
    d. taken apart and cleaned or replaced

3. Solutions used to degrease engine parts are commonly
    a. alkaline
    b. acidic
    c. neutral
    d. either alkaline or acidic

4. The ferrous hot tank solution is intended for cleaning parts made of
    a. ferrite
    b. aluminum-magnesium alloys
    c. stellite
    d. iron, steel

5. Cam bearings will be destroyed if left in the block during
    a. hot tanking
    b. steam cleaning
    c. washing in solvent
    d. machining

6. Solutions used to de-scale engine parts are commonly
    a. alkaline
    b. acidic

c. basic
d. neutral

7. Hot tank cleaning of working assemblies, such as pistons and rods, will cause
   a. solution contamination
   b. binding of moving parts
   c. no problems if done in non-ferrous solutions
   d. no problems if done in ferrous solutions

8. The bead blaster uses ________ for abrasive
   a. glass
   b. sand
   c. glass or sand
   d. silicon carbide

9. Flare brushes are used for
   a. cleaning threads
   b. cleaning ring grooves on pistons
   c. degreasing cylinder head ports
   d. decarbonizing combustion chambers

10. Particular care must be used when cleaning pistons not to
    a. use solvent
    b. use wire buffers
    c. use non-ferrous cleaning solution
    d. widen ring grooves

*Chapter 5, Inspecting Valve Train Components*

1. The limits of valve "rock" specifications will
   a. be equal to valve stem to guide clearance specifications
   b. be less than specified valve stem to guide clearance
   c. exceed specified valve stem to guide clearance
   d. have nothing to do with valve guides

2. The minimum margin acceptable after refacing valves is:
   a. 1/64" or one half of new thickness
   b. 1/32" or one half of new thickness
   c. 1/16"
   d. 3/32"

3. Wear on valve stems should not vary more than ________ from specified diameters
   a. .0005"

   b. .001″
   c. .0015″
   d. .002″

4. Valves should be visually checked for
   a. necking of the stem
   b. worn valve keeper grooves
   c. heat damage to valve faces and margins
   d. all of the above
   e. none of the above

5. Valve rotator action can be checked visually by inspecting wear on
   a. valve seats and valve faces
   b. valve guides
   c. valve stem tips
   d. rocker arm faces

6. Inspecting valve faces and margins for heat damage is critical because approximately ________ percent of heat transfer for cooling occurs through valve seats
   a. 76
   b. 30
   c. 24
   d. 15

7. Valve springs should be within ________ of specified tension in order to be re-used
   a. 10 lbs.
   b. 10%
   c. 5 lbs.
   d. 5%

8. Valve springs should be within ________ of being square for each 2″ in length
   a. 1/64″
   b. 1/32″
   c. 1/16″
   d. 3/32″

9. A new valve lifter base is shaped
   a. concave about .002″
   b. convex about .002″
   c. flat within .0002″
   d. flat within .000050″

10. A normal pattern of camshaft wear in an OHV engine would appear as
   a. a pattern extending full width of the rocker arm face
   b. a narrow pattern around each lobe

c. a pattern primarily to both edges of each lobe
d. a rounding of the cam lobe accompanied by pitting

11. Pre-determined leakdown will be excessive if
    a. plunger to body clearance is high, or a check valve malfunctions
    b. valve stem height is too short
    c. valve lifter bore clearance is high
    d. push rods are too short

12. The maximum allowable backlash of timing gears is
    a. plus or minus 1/4″
    b. .006″
    c. 10°
    d. 5°

13. The maximum allowable timing chain slack is
    a. plus or minus 1/4″
    b. .006″
    c. 10°
    d. 5°

14. Wear on the faces of rocker arms will
    a. cause valve lash or clearance adjustments to be loose
    b. ruin feeler gauges
    c. increase wear on valve stem faces
    d. any of the above
    e. none of the above

15. Stamped steel rocker arms should be ________ during valve grinds or engine overhauls
    a. refaced
    b. reconditioned
    c. replaced as a set
    d. replaced as required

16. Rocker arm studs should be inspected for
    a. wear on the sides and height above the head
    b. diameter and wear on the sides
    c. taper and height above the head
    d. diameter and height above the head

17. Solid push rods should be inspected for
    a. length
    b. straightness

c. end wear
d. length, straightness, and end wear

18. The following checks should be made for tubular push rods
    a. length
    b. end wear
    c. clear oil holes, if used
    d. all of above
    e. none of above

19. Camshaft binding in OHC cylinder heads may be caused by
    a. cam bearing wear
    b. cylinder head warpage
    c. cylinder head cracking
    d. severe corrosion

20. Cylinder head and engine block gasket surfaces should be checked for
    a. flatness
    b. burned or corroded areas
    c. nicks, scratches and dents
    d. all of the above
    e. none of the above

*Chapter 6, Inspecting Engine Block Components*

1. Cylinder taper as measured should not exceed ________ for new rings to seal properly
    a. .006″
    b. .008″
    c. .010″
    d. .012″

2. A piston skirt diameter measuring .005″ larger near the crown than near the ends of the skirts is
    a. collapsed
    b. expanded
    c. distorted
    d. slightly worn

3. Piston clearance is the difference between
    a. maximum cylinder diameter and minimum piston diameter
    b. minimum cylinder diameter and maximum piston diameter
    c. piston crown diameter and cylinder diameter below the ring travel
    d. piston skirt diameter and cylinder diameter at the points of maximum wear

4. The ring side clearance should not exceed ________ without recutting the ring groove
   a. .002″
   b. .004″
   c. .006″
   d. specified wear limits

5. A piston skirt measuring .005″ greater in diameter on a line 45° across the piston pin centerline than at any other point is
   a. collapsed
   b. expanded
   c. distorted
   d. slightly worn

6. As a general rule, cylinder blocks should be flat within
   a. .002″
   b. .002″ per foot
   c. .004″
   d. .004″ per foot

7. Gauging main bearing housing bore diameters will detect
   a. stretching of main bearing caps
   b. block warpage
   c. wear from running a bent crankshaft
   d. all of the above
   e. none of the above

8. Taper of a crankpin, or journal, is found by comparing measurements taken at
   a. points near each of the fillets
   b. three or more points around the circumference of the journal
   c. crankpins at opposite ends of the shaft
   d. journals at opposite ends of the shaft

9. Crankshaft straightness can be checked in the block by
   a. using plastigage at several points around the center journal
   b. removing the center main cap and using a dial indicator on the journal
   c. removing all main caps and using a dial indicator on the center journal
   d. removing all main caps and all but the end main bearings and using a dial indicator on the center journal

10. In operation, connecting rod housing bores
    a. wear oversize
    b. become tapered according to the crankpin taper

c. stretch in a direction 90° to the parting faces of the rod cap
d. stretch in a direction parallel to the parting faces of the rod cap

*Chapter 7, Crack Detection*

1. Magnetic particles collect
   a. in cracks
   b. in magnetic lines of force which surface at a crack
   c. along all lines of force between the magnetic poles, regardless of cracks
   d. in cracks or holes
2. Magnetic particle testing is limited to cracks in
   a. combustion chambers
   b. combustion chambers and ports
   c. combustion chambers, ports, and main bearing webs
   d. places which are visible
3. Wet magnaflux methods are generally used in automotive machine shops for
   a. cylinder heads
   b. engine blocks
   c. connecting rods and crankshafts
   d. all of the above
   e. none of the above
4. Pressure testing has advantages over other methods of checking cylinder heads or blocks because
   a. it can be used for magnetic castings
   b. it can be used for non-magnetic castings
   c. cracks need not be in visible areas
   d. all of the above
   e. none of the above
5. The correct sequence of steps for using dye penetrants is to apply
   a. penetrant, remove excess, apply developer
   b. developer, remove excess, apply penetrant
   c. a mix of penetrant and developer
   d. penetrant and wait

*Chapter 8, Crack Repair*

1. Crack repair should begin by tap drilling
   a. at one end of the crack
   b. 1/8″ past one end of the crack

c. at both ends of the crack
d. 1/8" past both ends of the crack

2. In order to repair a crack with pins, it is necessary that
   a. the full length of the crack can be reached with tools
   b. a replacement head be unavailable
   c. a seat insert is available for the head
   d. a guarantee not be required

3. The full length of cracks in combustion chambers need not be pinned if
   a. one end is "stop drilled"
   b. ceramic sealer is used
   c. the crack does not extend across valve seats
   d. pins from each end of the crack intersect within the casting

4. The most successful crack repairs can be made across valve seats by
   a. installing unthreaded taper pins
   b. peening the edges of the crack
   c. stop drilling
   d. installing pins and a seat insert

5. Ceramic seal circulators force ceramic sealer into cracks with
   a. cooling system pressure
   b. approximately 55 PSI
   c. approximately 110 PSI line pressure
   d. little or no pressure as the radiator cap is left loose

*Chapter 9, Reconditioning Valve Train Components*

1. By threading a capscrew into a valve guide, removal is done by
   a. using a puller on the capscrew
   b. driving against the guide
   c. using a puller on the guide
   d. driving against the capscrew

2. Before removing valve guides, the following check should be made
   a. height of valve guides above the spring seat
   b. valve stem diameter
   c. valve guide diameter
   d. valve guide to stem clearance

3. A knurling arbor is a tool which
   a. cuts threads
   b. forms threads

c. finishes valve guides to size
d. finishes valve guides to an oversize

**4.** In knurling valve guides, final clearance is obtained by
a. reaming
b. drilling
c. honing
d. knurling

**5.** In order to obtain original equipment life from knurled valve guides, guide wear should not exceed ________ prior to knurling
a. specified clearance limits
b. 002″
c. 004″
d. 006″

**6.** When using valve stems beyond the first oversize, it is recommended that
a. guides be reamed in .003″-.005″ increments up to final diameter
b. guides be reamed to size in a single pass with an oversize reamer of the correct diameter
c. guides be honed to size
d. false guides be fitted

**7.** One advantage to fitting "false" guides is
a. easier repair in the future
b. faster repair
c. cheaper repair
d. greater longevity

**8.** Valve seats are generally ground at
a. the same angle as valve faces
b. 1° or 2° less than the valve face
c. 1° or 2° more than the valve face
d. 44° to 45°

**9.** The minimum margin acceptable after refacing valves is
a. 1/64″ or one half of new thickness
b. 1/32″ or one half of new thickness
c. 1/16″
d. 3/32″

**10.** The general recommendation for the width of passenger car valve seats is
a. 1/32″ - 1/16″
b. 1/16″ - 3/32″
c. 3/32″ - 1/8″
d. 1/8″ - 3/16″

11. When grinding a typical 45° seat, the 30° stone will
    a. narrow the seat and move it closer to the margin
    b. narrow the seat and move it away from the margin
    c. widen the seat and move it closer to the margin
    d. widen the seat and move it away from the margin

12. Valve spring shims should be used to correct valve
    a. stem installed height
    b. spring installed height
    c. spring free length
    d. spring tension that is low

13. Lapping valve seats is recommended
    a. instead of seat grinding
    b. after all seat grinding
    c. before all seat grinding
    d. only when suitable surface finishes cannot be obtained by grinding or cutting

14. It is recommended that seat inserts be installed when
    a. installed spring height exceeds specs by more than .030″
    b. installed spring height exceeds specs by more than .060″
    c. valve stem height is .030″ over specs
    d. valve stem height is .060″ over specs

15. Welding around the inside of a seat insert will cause the seat to
    a. expand
    b. shrink
    c. break
    d. distort and break

16. After installing a seat, seat grinding is begun by grinding to
    a. the correct width
    b. the correct position on the valve face
    c. the correct depth for installed spring height
    d. uniform width, correct width, and correct position

17. An "O" ring type of valve seal is generally installed
    a. inside the spring retainer above the valve keepers
    b. below the spring retainer
    c. above the valve guide
    d. inside the spring retainer below the valve keepers

18. Positive valve seals
    a. clamp to guide

b. remain fixed to the stem
c. fit in a groove under the valve keepers
d. fit in a groove above the valve keepers

19. Rocker arm studs are repaired by removing old studs and
    a. press fitting standard size studs
    b. press fitting .001″ oversize studs
    c. reaming first, then press fitting .001″ oversize studs
    d. reaming first, then press fitting .003″ oversize studs

20. Failure to reduce the installed height of valve stems to specifications may cause valves
    a. to remain open
    b. to remain closed
    c. to burn
    d. lifter noise

*Chapter 10, Reconditioning Engine Block Components*

1. Special fine grit stones are required for finish honing when using
    a. chrome oil rings
    b. iron oxide coated compression rings
    c. chrome plated compression rings
    d. moly coated compression rings

2. Too fine a surface finish on cylinders will cause
    a. rapid wear of chrome rings
    b. chrome rings not to seat
    c. iron rings not to seat
    d. rapid wear of iron rings

3. A glaze-breaking cylinder hone will
    a. produce a round bore
    b. straighten a tapered cylinder
    c. will polish the cylinder finish
    d. will roughen the cylinder finish

4. Cylinders should be finish bored a minimum of ________ undersize for hand honing.
    a. .001″ – .002″
    b. .002″ – .003″
    c. .003″ – .005″
    d. .005″ – .007″

5. Cylinder alignment and location when boring is determined by the
    a. original cylinder centerline

b. the centerline of the boring bar spindle
c. crankshaft centerline
d. final honing operation

6. The purpose of a cylinder chamfer is to
   a. prevent cutting yourself on sharp edges
   b. prevent detonation during combustion
   c. allow rings to enter cylinders easily
   d. allow rings to enter cylinders without using a ring compressor

7. Cylinder sleeves may be installed by
   a. "shrink" fitting
   b. press fitting
   c. driving them in
   d. any of the above
   e. none of the above

8. An engine is fitted with different main caps. The easiest way of correcting the crankshaft centerline is to
   a. get another block
   b. line bore
   c. line hone
   d. line bore or hone, no difference

9. Thrust faces of flanged bearings can
   a. be refaced in a line hone
   b. be refaced in a lining boring machine
   c. be refaced only in a lathe
   d. not be refaced

10. Align boring or honing causes the housing bore centerline to
    a. move further from the camshaft centerline
    b. move closer to the camshaft centerline
    c. locate in the original position
    d. remain unchanged

11. When grinding main caps, the parting face of the thrust main cap must be ground
    a. 90° to the thrust face
    b. 90° the fixed jaw of the rod and cap grinder
    c. in the same plane as the original parting face
    d. so that the housing bore is made undersize for machining without consideration of the angle or position

12. Piston ring side clearance must be less than
    a. .001″
    b. .004″
    c. .006″
    d. .008″

13. Oil burning due to "oil pumping" is due to excess
    a. oil pressure
    b. ring groove side clearance
    c. rod bearing clearance
    d. rod side clearance

14. If the piston is in serviceable condition but piston clearance is excessive, piston fit can be improved by
    a. installing ring groove inserts
    b. knurling the piston
    c. honing the cylinder
    d. removing the ring ridge

15. A connecting rod with a pin bore too large for an intended press fit should be
    a. bushed to standard size
    b. fitted with oversize pins or replaced
    c. honed to size
    d. bushed and honed to standard size

16. The procedure for replacing bushings in rods for full floating pins is to press fit new bushings
    a. of correct I.D.
    b. and hone to size
    c. and expand in place
    d. and expand in place and hone to size

17. When press-fitting piston pins through connecting rods, take this precaution
    a. heat the piston pin first
    b. use .002″ interference
    c. press from one side only
    d. press back and forth from each direction until the pin is centered

18. Stock removal from each parting face of connecting rods should be a minimum of ________ before resizing housing bores.
    a. .001″
    b. .002″
    c. .003″
    d. .004″

19. Rod housing bores are resized to the
    a. next oversize

b. specified oil clearance
c. specified diameter
d. specified diameter plus the bearing oversize

20. Rods must be straight and free of twist within
a. .0006″ in 1″
b. .001″ in 6″
c. .001″
d. .006″

21. When aligning rods
a. bend is corrected first
b. twist is corrected first
c. bend or twist may be corrected first
d. housing bores must be resized first

22. When cam lobes are reground
a. they are built up by welding first
b. lift is reduced
c. duration is reduced
d. the base circle diameter is reduced and lift is unchanged

23. The following valve lifters are refaced and reused:
a. hydraulics
b. solids
c. hard lifters
d. soft lifters

24. Bent crankshafts are
a. reground straight
b. straightened and reused
c. straightened before regrinding
d. replaced

25. A crankpin or journal may be welded before grinding so that
a. cracks are repaired
b. the shaft is reground to standard
c. stroke length may be corrected
d. other crankpins or journals need not be ground to greater undersizes

26. The required surface finish quality on crankshafts is obtained by
a. grinding
b. grinding and polishing
c. nitriding
d. chrome plating

27. In order to minimize the possibility of flywheel run-out, flywheels should be refaced
    a. parallel to the original clutch surface
    b. parallel the reverse side where the flywheel is bolted to the crankshaft
    c. the minimum amount required for clean-up
    d. all of the above
    e. none of the above

28. The amount of stock removal permitted for resurfacing flywheels is frequently
    a. .010″
    b. .020″
    c. .030″
    d. not specified

29. When clearance between oil pump gears and the pump housing is excessive, the pump should be
    a. fitted with new gears
    b. fitted with oversize gears
    c. remachined
    d. replaced

30. Oil pump pick-ups which do not fit tightly into the pump may
    a. cause oil to become aerated
    b. fall-out
    c. cause engine failure due to poor lubrication or no lubrication
    d. any of the above
    e. none of the above

*Chapter 11, Resurfacing Cylinder Heads and Blocks*

1. Failure to permit a grinding machine to "spark out" will cause the
    a. grinding wheel and work to overheat
    b. grinding wheel to become glazed
    c. surface not to be flat
    d. surface finish to be unsatisfactory

2. The advantage to milling cutters over grinding wheels is
    a. faster stock removal
    b. a higher quality surface finish
    c. improved performance when resurfacing blocks or heads with hard spots
    d. any of above
    e. none of above

3. When testing for flatness with a straight edge and thickness gauge, the recommended limit for surface deviation is:
    a. .002″

b. .004″
c. .006″
d. .008″

4. When too much resurfacing is done on blocks or heads, problems may be encountered with:
   a. piston to cylinder head interference
   b. valve to piston interference
   c. non-adjustable rocker arms holding valves open
   d. any of the above
   e. none of the above

5. Some effects of improper manifold, block, and cylinder head resurfacing on V-block engines could include:
   a. vacuum and oil leaks
   b. power loss
   c. oil burning
   d. any of the above
   e. none of the above

6. After resurfacing V-block cylinder heads .020″, alignments are corrected by
   a. using thicker head gaskets
   b. machining the top of the block
   c. machining the intake manifold side
   d. no single step above

7. OHC cylinder head warpage, and required resurfacing, can cause:
   a. valve timing changes
   b. cam bearing alignment changes
   c. timing chain slack to increase
   d. any of above
   e. none of the above

8. Composition gaskets may be used in place of embossed steel gaskets to compensate for approximately ________ of resurfacing.
   a. .000″
   b. .010″
   c. .020″
   d. .040″

9. An engine has a 90° V-block and each cylinder head has a 90° included angle. The cylinder heads were surfaced .020″, therefore ________ should surface from the intake side of the cylinder heads.
   a. .020″

b. .017″
c. .012″
d. .010″

10. An engine has a 90° V-block and each cylinder head has a 80° included angle. The cylinder heads were surfaced .010″, therefore ________ should be surfaced from the top of the block.
    a. .015″
    b. .017″
    c. .019″
    d. .021″

*Chapter 12, Engine Balancing*

1. An unbalance condition in a crankshaft
    a. is greatest at idle
    b. is greatest when under load
    c. is greatest when decelerating
    d. increases in force with increases in RPM
2. Pistons are considered to be ________ weight
    a. reciprocating
    b. rotating
    c. dynamic
    d. static
3. The reciprocating weights of a set of connecting rods are equalized. Variations in total weight are due to
    a. variations in reciprocating weight before balancing
    b. variations in rotating weight
    c. errors in rod center-distance
    d. errors in weighing
4. Bob-weights for a V-eight crankshaft are made up of
    a. all reciprocating weight and one-half of rotating weight
    b. all rotating weight and one-half of reciprocating weight
    c. all rotating weight and all reciprocating weight
    d. one-half of all rotating and reciprocating weight
5. An "externally weighted" engine is balanced with the flywheel and the harmonic balancer done
    a. separately from the crankshaft
    b. with the crankshaft
    c. before the crankshaft
    d. separately from each other

1. Prior to assembly, threaded head bolt holes should be:
   a. cleaned and chamfered
   b. lubricated
   c. chamfered
   d. cleaned, chamfered, and lubricated

2. Valve spring compressors are
   a. adjusted to clear valve stems
   b. adjusted to clear valve seals
   c. adjusted to clear valve springs
   d. not adjusted

3. A core hole in an engine block measures 1-1/2″ in diameter. The core plug for this hole will be ________ in diameter.
   a. under 1-1/2″
   b. exactly 1-1/2″
   c. over 1-1/2″
   d. 1-5/8″

4. Oil plugs are installed
   a. dry
   b. with gasket sealer around the outer edges
   c. with gasket sealer on the back side
   d. with gasket sealer on all sides

5. Oil holes drilled in cam bearings are intended to provide oil for:
   a. cam journals
   b. rocker arm assemblies
   c. hydraulic lifters
   d. any of the above, depending upon the engine
   e. none of the above, regardless of the engine

6. A camshaft which binds in new cam bearings is made to rotate freely by:
   a. going ahead with engine assembly and allowing the bearings to "run-in"
   b. replacing the tight bearing
   c. removing tight bearings, rotating them 180°, and re-installing them
   d. removing high spots with a hand scraper

7. The reason for coating a camshaft with anti-scuff lube is to:
   a. lubricate the camshaft until the oil pump is primed
   b. prevent binding of camshaft in new bearings
   c. prevent valve lifters which are not primed from damaging cam lobes
   d. prevent scuffing during initial break-in

8. High camshaft end play is corrected by replacing the
   a. camshaft sprocket
   b. spacer ring
   c. thrust plate
   d. camshaft

9. Lip seals on rotating shafts should be:
   a. left dry for break-in
   b. lubricated for break-in
   c. lubricated only if it is required in service manuals
   d. lubricated if it is a timing cover seal and left dry if it is a rearmain seal

10. Rear main seals are shaped with:
    a. a driver the same diameter as the rear main
    b. a driver the same diameter as the seal surface of the crankshaft
    c. round stock
    d. a or b above
    e. b or c above

11. Absorbent side seals should be:
    a. lubricated only to aid in assembly
    b. installed dry and then oiled
    c. oil soaked and installed before swelling begins
    d. installed dry to prevent swelling

12. The thrust main bearing:
    a. varies in location
    b. is the rear main bearing
    c. is the center main bearing
    d. is the front main bearing

13. When installing the crankshaft in the block, it should be checked for:
    a. binding
    b. oil clearance
    c. end play
    d. all of above
    e. none of above

14. Valve timing is set during engine assembly by
    a. measuring valve overlap on the intake stroke
    b. measuring valve opening on the exhaust stroke
    c. lining up the camshaft so both valves are closed on the power stroke
    d. lining up camshaft and crankshaft timing marks according to specifications

15. Minimum piston ring end gap is approximately
    a. .003″ per inch of cylinder diameter

b. .003″
c. .009″
d. .009-.012″

16. A cylinder is honed .003″ oversize. The ring end gap will be
    a. unchanged
    b. decreased .003″
    c. increased .009″
    d. .003″ per inch of cylinder diameter

17. A ring compressor is used to
    a. fit rings on pistons
    b. fit pistons and rings into cylinders
    c. guide pistons into cylinders
    d. reduce ring end gaps

18. Excessive rod side clearance may be corrected by
    a. re-sizing the rods
    b. grinding the crankshaft
    c. replacing the rod bearings
    d. replacing the rods

19. It is recommended that nuts or bolts for bearing caps be torqued as follows
    a. "dry"
    b. "wet"
    c. lubricated with engine oil and torqued to the mid-range of specifications
    d. lubricated with anti-seize lube and torqued to the mid-range of specifications

20. Tightening sequences for head bolts generally:
    a. start at one end of the head and work across
    b. start in the middle and work alternately left and right around each cylinder
    c. follow no particular pattern
    d. any of above depending upon the engine
    e. none of above

21. When gasket sealer is used in head gasket replacement, the sealer should be
    a. applied to the head surfaces
    b. the hardening type
    c. applied to the gasket
    d. applied to the block and head surfaces

22. Rocker arm assemblies must be checked for
    a. right and left offsets of rocker arms
    b. direction or position of oil holes in rocker shafts

c. plugs in the ends of rocker shafts
d. any of above depending on engine design
e. none of above regardless of engine design

23. When making valve and rocker arm adjustments on an OHV engine with hydraulic valve lifters, adjusting nuts are tightened to
    a. zero clearance and backed off 1/2 to 3/4 of a turn
    b. zero clearance
    c. zero clearance and tightened 1/2 to 3/4 of a turn
    d. the specified valve clearance

24. The oil pump pick-up should be checked to see that it is:
    a. tight in the pump body
    b. within at least 1/4" of the bottom of the oil pan
    c. clean
    d. all of above
    e. none of above

25. Newly assembled engines may be pre-oiled before starting by
    a. assembling parts with engine oil
    b. turning the oil pump before installing the distributor
    c. connecting a pre-lubricator to the engine
    d. either a or b above
    e. either b or c above

*Chapter 14, Engine Installation and Break-In*

1. An engine should be broken in on:
    a. inexpensive bulk oil
    b. non-detergent oil
    c. an oil of service rating specified for the engine
    d. special break-in oil

2. Using friction-reducing oil additives in a new engine will likely:
    a. cause no harm
    b. extend engine life
    c. stop camshaft wear
    d. inhibit ring seating

3. Piston rings are best seated by
    a. driving at constant speeds
    b. driving at low speeds
    c. driving at high speeds
    d. accelerating and decelerating through several cycles

4. Engine bearings ________ during break-in.
   a. wear in like piston rings
   b. conform to shape and harden
   c. become softer
   d. absorb particles

5. Service after break-in should include
   a. rechecking valve adjustment on all engines without hydraulic valve lifters
   b. changing engine oil and filter
   c. retightening head bolts
   d. all of above
   e. none of above

# Appendix A

# Reference Tables

METRIC-ENGLISH CONVERSION TABLE

| mm | Inches | mm | Inches | mm | Inches | mm | Inches | mm | Inches |
|---|---|---|---|---|---|---|---|---|---|
| 0.01 | .00039 | 0.41 | .01614 | 0.81 | .03189 | 21 | .82677 | 61 | 2.40157 |
| 0.02 | .00079 | 0.42 | .01654 | 0.82 | .03228 | 22 | .86614 | 62 | 2.44094 |
| 0.03 | .00118 | 0.43 | .01693 | 0.83 | .03268 | 23 | .90551 | 63 | 2.48031 |
| 0.04 | .00157 | 0.44 | .01732 | 0.84 | .03307 | 24 | .94488 | 64 | 2.51968 |
| 0.05 | .00197 | 0.45 | .01772 | 0.85 | .03346 | 25 | .98425 | 65 | 2.55905 |
| 0.06 | .00236 | 0.46 | .01811 | 0.86 | .03386 | 26 | 1.02362 | 66 | 2.59842 |
| 0.07 | .00276 | 0.47 | .01850 | 0.87 | .03425 | 27 | 1.06299 | 67 | 2.63779 |
| 0.08 | .00315 | 0.48 | .01890 | 0.88 | .03465 | 28 | 1.10236 | 68 | 2.67716 |
| 0.09 | .00354 | 0.49 | .01929 | 0.89 | .03504 | 29 | 1.14173 | 69 | 2.71653 |
| 0.10 | .00394 | 0.50 | .01969 | 0.90 | .03543 | 30 | 1.18110 | 70 | 2.75590 |
| 0.11 | .00433 | 0.51 | .02008 | 0.91 | .03583 | 31 | 1.22047 | 71 | 2.79527 |
| 0.12 | .00472 | 0.52 | .02047 | 0.92 | .03622 | 32 | 1.25984 | 72 | 2.83464 |
| 0.13 | .00512 | 0.53 | .02087 | 0.93 | .03661 | 33 | 1.29921 | 73 | 2.87401 |
| 0.14 | .00551 | 0.54 | .02126 | 0.94 | .03701 | 34 | 1.33858 | 74 | 2.91338 |
| 0.15 | .00591 | 0.55 | .02165 | 0.95 | .03740 | 35 | 1.37795 | 75 | 2.95275 |
| 0.16 | .00630 | 0.56 | .02205 | 0.96 | .03780 | 36 | 1.41732 | 76 | 2.99212 |
| 0.17 | .00669 | 0.57 | .02244 | 0.97 | .03819 | 37 | 1.45669 | 77 | 3.03149 |
| 0.18 | .00709 | 0.58 | .02283 | 0.98 | .03858 | 38 | 1.49606 | 78 | 3.07086 |
| 0.19 | .00748 | 0.59 | .02323 | 0.99 | .03898 | 39 | 1.53543 | 79 | 3.11023 |
| 0.20 | .00787 | 0.60 | .02362 | 1.00 | .03937 | 40 | 1.57480 | 80 | 3.14960 |
| 0.21 | .00827 | 0.61 | .02402 | 1 | .03937 | 41 | 1.61417 | 81 | 3.18897 |
| 0.22 | .00866 | 0.62 | .02441 | 2 | .07874 | 42 | 1.65354 | 82 | 3.22834 |
| 0.23 | .00906 | 0.63 | .02480 | 3 | .11811 | 43 | 1.69291 | 83 | 3.26771 |
| 0.24 | .00945 | 0.64 | .02520 | 4 | .15748 | 44 | 1.73228 | 84 | 3.30708 |
| 0.25 | .00984 | 0.65 | .02559 | 5 | .19685 | 45 | 1.77165 | 85 | 3.34645 |
| 0.26 | .01024 | 0.66 | .02598 | 6 | .23622 | 46 | 1.81102 | 86 | 3.38582 |
| 0.27 | .01063 | 0.67 | .02638 | 7 | .27559 | 47 | 1.85039 | 87 | 3.42519 |
| 0.28 | .01102 | 0.68 | .02677 | 8 | .31496 | 48 | 1.88976 | 88 | 3.46456 |
| 0.29 | .01142 | 0.69 | .02717 | 9 | .35433 | 49 | 1.92913 | 89 | 3.50393 |
| 0.30 | .01181 | 0.70 | .02756 | 10 | .39370 | 50 | 1.96850 | 90 | 3.54330 |
| 0.31 | .01220 | 0.71 | .02795 | 11 | .43307 | 51 | 2.00787 | 91 | 3.58267 |
| 0.32 | .01260 | 0.72 | .02835 | 12 | .47244 | 52 | 2.04724 | 92 | 3.62204 |
| 0.33 | .01299 | 0.73 | .02874 | 13 | .51181 | 53 | 2.08661 | 93 | 3.66141 |
| 0.34 | .01339 | 0.74 | .02913 | 14 | .55118 | 54 | 2.12598 | 94 | 3.70078 |
| 0.35 | .01378 | 0.75 | .02953 | 15 | .59055 | 55 | 2.16535 | 95 | 3.74015 |
| 0.36 | .01417 | 0.76 | .02992 | 16 | .62992 | 56 | 2.20472 | 96 | 3.77952 |
| 0.37 | .01457 | 0.77 | .03032 | 17 | .66929 | 57 | 2.24409 | 97 | 3.81889 |
| 0.38 | .01496 | 0.78 | .03071 | 18 | .70866 | 58 | 2.28346 | 98 | 3.85826 |
| 0.39 | .01535 | 0.79 | .03110 | 19 | .74803 | 59 | 2.32283 | 99 | 3.89763 |
| 0.40 | .01575 | 0.80 | .03150 | 20 | .78740 | 60 | 2.36220 | 100 | 3.93700 |

## ENGLISH-METRIC CONVERSION TABLE

| Inches Dec. | mm | Inches Dec. | mm |
|---|---|---|---|
| 0.01 | 0.2540 | 0.51 | 12.9540 |
| 0.02 | 0.5080 | 0.52 | 13.2080 |
| 0.03 | 0.7620 | 0.53 | 13.4620 |
| 0.04 | 1.0160 | 0.54 | 13.7160 |
| 0.05 | 1.2700 | 0.55 | 13.9700 |
| 0.06 | 1.5240 | 0.56 | 14.2240 |
| 0.07 | 1.7780 | 0.57 | 14.4780 |
| 0.08 | 2.0320 | 0.58 | 14.7320 |
| 0.09 | 2.2860 | 0.59 | 14.9860 |
| 0.10 | 2.5400 | 0.60 | 15.2400 |
| 0.11 | 2.7940 | 0.61 | 15.4940 |
| 0.12 | 3.0480 | 0.62 | 15.7480 |
| 0.13 | 3.3020 | 0.63 | 16.0020 |
| 0.14 | 3.5560 | 0.64 | 16.2560 |
| 0.15 | 3.8100 | 0.65 | 16.5100 |
| 0.16 | 4.0640 | 0.66 | 16.7640 |
| 0.17 | 4.3180 | 0.67 | 17.0180 |
| 0.18 | 4.5720 | 0.68 | 17.2720 |
| 0.19 | 4.8260 | 0.69 | 17.5260 |
| 0.20 | 5.0800 | 0.70 | 17.7800 |
| 0.21 | 5.3340 | 0.71 | 18.0340 |
| 0.22 | 5.5880 | 0.72 | 18.2880 |
| 0.23 | 5.8420 | 0.73 | 18.5420 |
| 0.24 | 6.0960 | 0.74 | 18.7960 |
| 0.25 | 6.3500 | 0.75 | 19.0500 |
| 0.26 | 6.6040 | 0.76 | 19.3040 |
| 0.27 | 6.8580 | 0.77 | 19.5580 |
| 0.28 | 7.1120 | 0.78 | 19.8120 |
| 0.29 | 7.3660 | 0.79 | 20.0660 |
| 0.30 | 7.6200 | 0.80 | 20.3200 |
| 0.31 | 7.8740 | 0.81 | 20.5740 |
| 0.32 | 8.1280 | 0.82 | 20.8280 |
| 0.33 | 8.3820 | 0.83 | 21.0820 |
| 0.34 | 8.6360 | 0.84 | 21.3360 |
| 0.35 | 8.8900 | 0.85 | 21.5900 |
| 0.36 | 9.1440 | 0.86 | 21.8440 |
| 0.37 | 9.3980 | 0.87 | 22.0980 |
| 0.38 | 9.6520 | 0.88 | 22.3520 |
| 0.39 | 9.9060 | 0.89 | 22.6060 |
| 0.40 | 10.1600 | 0.90 | 22.8600 |
| 0.41 | 10.4140 | 0.91 | 23.1140 |
| 0.42 | 10.6680 | 0.92 | 23.3680 |
| 0.43 | 10.9220 | 0.93 | 23.6220 |
| 0.44 | 11.1760 | 0.94 | 23.8760 |
| 0.45 | 11.4300 | 0.95 | 24.1300 |
| 0.46 | 11.6840 | 0.96 | 24.3840 |
| 0.47 | 11.9380 | 0.97 | 24.6380 |
| 0.48 | 12.1920 | 0.98 | 24.8920 |
| 0.49 | 12.4460 | 0.99 | 25.1460 |
| 0.50 | 12.7000 | 1.00 | 25.4000 |

| Inches Frac. | Inches Dec. | mm | Inches Frac. | Inches Dec. | mm |
|---|---|---|---|---|---|
| 1/64 | 0.015625 | 0.3969 | 33/64 | 0.515625 | 13.0969 |
| 1/32 | 0.031250 | 0.7938 | 17/32 | 0.531250 | 13.4938 |
| 3/64 | 0.046875 | 1.1906 | 35/64 | 0.546875 | 13.8906 |
| 1/16 | 0.062500 | 1.5875 | 9/16 | 0.562500 | 14.2875 |
| 5/64 | 0.078125 | 1.9844 | 37/64 | 0.578125 | 14.6844 |
| 3/32 | 0.093750 | 2.3812 | 19/32 | 0.593750 | 15.0812 |
| 7/64 | 0.109375 | 2.7781 | 39/64 | 0.609375 | 15.4781 |
| 1/8 | 0.125000 | 3.1750 | 5/8 | 0.625000 | 15.8750 |
| 9/64 | 0.140625 | 3.5719 | 41/64 | 0.640625 | 16.2719 |
| 5/32 | 0.156250 | 3.9688 | 21/32 | 0.656250 | 16.6688 |
| 11/64 | 0.171875 | 4.3656 | 43/64 | 0.671875 | 17.0656 |
| 3/16 | 0.187500 | 4.7625 | 11/16 | 0.687500 | 17.4625 |
| 13/64 | 0.203125 | 5.1594 | 45/64 | 0.703125 | 17.8594 |
| 7/32 | 0.218750 | 5.5562 | 23/32 | 0.718750 | 18.2562 |
| 15/64 | 0.234375 | 5.9531 | 47/64 | 0.734375 | 18.6531 |
| 1/4 | 0.250000 | 6.3500 | 3/4 | 0.750000 | 19.0500 |
| 17/64 | 0.265625 | 6.7469 | 49/64 | 0.765625 | 19.4469 |
| 9/32 | 0.281250 | 7.1438 | 25/32 | 0.781250 | 19.8437 |
| 19/64 | 0.296875 | 7.5406 | 51/64 | 0.796875 | 20.2406 |
| 5/16 | 0.312500 | 7.9375 | 13/16 | 0.812500 | 20.6375 |
| 21/64 | 0.328125 | 8.3344 | 53/64 | 0.828125 | 21.0344 |
| 11/32 | 0.343750 | 8.7312 | 27/32 | 0.843750 | 21.4312 |
| 23/64 | 0.359375 | 9.1281 | 55/64 | 0.859375 | 21.8281 |
| 3/8 | 0.375000 | 9.5250 | 7/8 | 0.875000 | 22.2250 |
| 25/64 | 0.390625 | 9.9219 | 57/64 | 0.890625 | 22.6219 |
| 13/32 | 0.406250 | 10.3188 | 29/32 | 0.906250 | 23.0188 |
| 27/64 | 0.421875 | 10.7156 | 59/64 | 0.921875 | 23.4156 |
| 7/16 | 0.437500 | 11.1125 | 15/16 | 0.937500 | 23.8125 |
| 29/64 | 0.453125 | 11.5094 | 61/64 | 0.953125 | 24.2094 |
| 15/32 | 0.468750 | 11.9062 | 31/32 | 0.968750 | 24.6062 |
| 31/64 | 0.484375 | 12.3031 | 63/64 | 0.984375 | 25.0031 |
| 1/2 | 0.500000 | 12.7000 | 1 | 1.000000 | 25.4000 |

## DECIMAL EQUIVALENTS FOR FRACTIONAL, WIRE GAUGE AND LETTER-SIZE DRILLS

| Drill Size | Decimal | Drill Size | Decimal | Drill Size | Decimal | Drill Size | Decimal |
|---|---|---|---|---|---|---|---|
| 80 | .0135 | 42 | .0935 | 13/64 | .2031 | X | .3970 |
| 79 | .0145 | 3/32 | .0938 | 6 | .2040 | Y | .4040 |
| 1/64 | .0156 | 41 | .0960 | 5 | .2055 | 13/32 | .4062 |
| 78 | .0160 | 40 | .0980 | 4 | .2090 | Z | .4130 |
| 77 | .0180 | 39 | .0995 | 3 | .2130 | 27/64 | .4219 |
| 76 | .0200 | 38 | .1015 | 7/32 | .2188 | 7/16 | .4375 |
| 75 | .0210 | 37 | .1040 | 2 | .2210 | 29/64 | .4531 |
| 74 | .0225 | 36 | .1065 | 1 | .2280 | 15/32 | .4688 |
| 73 | .0240 | 7/64 | .1094 | A | .2340 | 31/64 | .4844 |
| 72 | .0250 | 35 | .1100 | 15/64 | .2344 | 1/2 | .5000 |
| 71 | .0260 | 34 | .1110 | B | .2380 | 33/64 | .5156 |
| 70 | .0280 | 33 | .1130 | C | .2420 | 17/32 | .5312 |
| 69 | .0292 | 32 | .1160 | D | .2460 | 35/64 | .5469 |
| 68 | .0310 | 31 | .1200 | 1/4 | .2500 | 9/16 | .5625 |
| 1/32 | .0312 | 1/8 | .1250 | E | .2500 | 37/64 | .5781 |
| 67 | .0320 | 30 | .1285 | F | .2570 | 19/32 | .5938 |
| 66 | .0330 | 29 | .1360 | G | .2610 | 39/64 | .6094 |
| 65 | .0350 | 28 | .1405 | 17/64 | .2656 | 5/8 | .6250 |
| 64 | .0360 | 9/64 | .1406 | H | .2660 | 41/64 | .6406 |
| 63 | .0370 | 27 | .1440 | I | .2720 | 21/32 | .6562 |
| 62 | .0380 | 26 | .1470 | J | .2770 | 43/64 | .6719 |
| 61 | .0390 | 25 | .1495 | K | .2810 | 11/16 | .6875 |
| 60 | .0400 | 24 | .1520 | 9/32 | .2812 | 45/64 | .7031 |
| 59 | .0410 | 23 | .1540 | L | .2900 | 23/32 | .7188 |
| 58 | .0420 | 5/32 | .1562 | M | .2950 | 47/64 | .7344 |
| 57 | .0430 | 22 | .1570 | 19/64 | .2969 | 3/4 | .7500 |
| 56 | .0465 | 21 | .1590 | N | .3020 | 49/64 | .7656 |
| 3/64 | .0469 | 20 | .1610 | 5/16 | .3125 | 25/32 | .7812 |
| 55 | .0520 | 19 | .1660 | O | .3160 | 51/64 | .7969 |
| 54 | .0550 | 18 | .1695 | P | .3230 | 13/16 | .8125 |
| 53 | .0595 | 11/64 | .1719 | 21/64 | .3281 | 53/64 | .8281 |
| 1/16 | .0625 | 17 | .1730 | Q | .3320 | 27/32 | .8438 |
| 52 | .0635 | 16 | .1770 | R | .3390 | 55/64 | .8594 |
| 51 | .0670 | 15 | .1800 | 11/32 | .3438 | 7/8 | .8750 |
| 50 | .0700 | 14 | .1820 | S | .3480 | 57/64 | .8906 |
| 49 | .0730 | 13 | .1850 | T | .3580 | 29/32 | .9062 |
| 48 | .0760 | 3/16 | .1875 | 23/64 | .3594 | 59/64 | .9219 |
| 5/64 | .0781 | 12 | .1890 | U | .3680 | 15/16 | .9375 |
| 47 | .0785 | 11 | .1910 | 3/8 | .3750 | 61/64 | .9531 |
| 46 | .0810 | 10 | .1935 | V | .3770 | 31/32 | .9688 |
| 45 | .0820 | 9 | .1960 | W | .3860 | 63/64 | .9844 |
| 44 | .0860 | 8 | .1990 | 25/64 | .3906 | 1 | 1.0000 |
| 43 | .0890 | 7 | .2010 | | | | |

## TAP DRILL SIZES
## BASED ON APPROXIMATELY 75% FULL THREAD

### National Coarse and Fine Threads

| Thread | Drill | Thread | Drill |
|---|---|---|---|
| 0 - 80 | 3/64 | 7/16 - 14 | U |
| 1 - 64 | 53 | 7/16 - 20 | 25/64 |
| 1 - 72 | 53 | 1/2 - 12 | 27/64 |
| 2 - 56 | 50 | 1/2 - 13 | 27/64 |
| 2 - 64 | 50 | 1/2 - 20 | 29/64 |
| 3 - 48 | 47 | 9/16 - 12 | 31/64 |
| 3 - 56 | 45 | 9/16 - 18 | 33/64 |
| 4 - 40 | 43 | 5/8 - 11 | 17/32 |
| 4 - 48 | 42 | 5/8 - 18 | 37/64 |
| 5 - 40 | 38 | 3/4 - 10 | 21/32 |
| 5 - 44 | 37 | 3/4 - 16 | 11/16 |
| 6 - 32 | 36 | 7/8 - 9 | 49/64 |
| 6 - 40 | 33 | 7/8 - 14 | 13/16 |
| 8 - 32 | 29 | 1 - 8 | 7/8 |
| 8 - 36 | 29 | 1 - 12 | 59/64 |
| 10 - 24 | 25 | 1 1/8 - 7 | 63/64 |
| 10 - 32 | 21 | 1 1/8 - 12 | 1 - 3/64 |
| 12 - 24 | 16 | 1 1/4 - 7 | 1 - 7/64 |
| 12 - 28 | 14 | 1 1/4 - 12 | 1 - 11/64 |
| 1/4 - 20 | 7 | 1 3/8 - 6 | 1 - 7/32 |
| 1/4 - 28 | 3 | 1 3/8 - 12 | 1 - 19/64 |
| 5/16 - 18 | F | 1 1/2 - 6 | 1 - 11/32 |
| 5/16 - 24 | I | 1 1/2 - 12 | 1 - 27/64 |
| 3/8 - 16 | 5/16 | 1 3/4 - 5 | 1 - 9/16 |
| 3/8 - 24 | Q | | |

### Taper Pipe

| Thread | Drill |
|---|---|
| 1/8 - 27 | R |
| 1/4 - 18 | 7/16 |
| 3/8 - 18 | 37/64 |
| 1/2 - 14 | 23/32 |
| 3/4 - 14 | 59/64 |
| 1 - 11 1/2 | 1 - 5/32 |
| 1 1/4 - 11 1/2 | 1 - 1/2 |
| 1 1/2 - 11 1/2 | 1 - 47/64 |
| 2 - 11 1/2 | 2 - 7/32 |
| 2 1/2 - 8 | 2 5/8 |
| 3 - 8 | 3 1/4 |
| 3 1/2 - 8 | 3 3/4 |
| 4 - 8 | 4 1/4 |

### Straight Pipe

| Thread | Drill |
|---|---|
| 1/8 - 27 | S |
| 1/4 - 18 | 29/64 |
| 3/8 - 18 | 19/32 |
| 1/2 - 14 | 47/64 |
| 3/4 - 14 | 15/16 |
| 1 - 11 1/2 | 1 - 3/16 |
| 1 1/4 - 11 1/2 | 1 - 33/64 |
| 1 1/2 - 11 1/2 | 1 - 3/4 |
| 2 - 11 1/2 | 2 - 7/32 |
| 2 1/2 - 8 | 2 - 21/32 |
| 3 - 8 | 3 - 9/32 |
| 3 1/2 - 8 | 3 - 25/32 |
| 4 - 8 | 4 - 9/32 |

## TORQUE AND GRADE SPECIFICATIONS

(Comparisons are based on requirements established by the Society of Automotive Engineers)

| Head Marking | Type of Steel | S.A.E. Grade | Tensile Strength (P.S.I.) | Tensile Strength of ¾" dia. Bolt U.S.S. |
|---|---|---|---|---|
| | 1018–1022 Low Carbon | 2 | 62,600 Avg. | 21,400 lbs. |
| | 1038 Medium Carbon Heat Treated | 5 | 113,300 Avg. | 40,100 lbs. |
| | 1041 Medium Carbon Heat Treated | 6* | 136,500 Avg. | 44,400 lbs. |
| | Medium Carbon 4037 Alloy Heat Treated | 8 | 150,000 | 50,000 lbs. |
| | BOWMALLOY Chrome Nickel Alloy Steel | No S.A.E. Rating High Enough | 186,820 | 62,460 lbs. |

*Previous SAE Designation Compliments of Bowman Products Division, Associated Spring Corporation

## RECOMMENDED ASSEMBLY TORQUES
## FOR BOWMAN HEX HEAD CAPSCREWS WITH BOWMALLOY NUTS

| | SAE 1020 Capscrews | | SAE 1038 Heat Treated Capscrews | | SAE 1041 Heat Treated Capscrews | | Bowmalloy | |
|---|---|---|---|---|---|---|---|---|
| | Bright Oil Finish | | SAE Grade 5 Plated | | SAE Grade 6* Plated | | Over SAE Grade 8 Plated | |
| | Recommended Torque (Foot-pounds) | | Recommended Torque (Foot-pounds) | | Recommended Torque (Foot-pounds) | | Recommended Torque (Foot-pounds) | |
| Dia. | UNC | UNF | UNC | UNF | UNC | UNF | UNC | UNF |
| 1/4 | 6 | 7 | 10 | 12 | 12 | 15 | 14 | 17 |
| 5/16 | 13 | 14 | 20 | 22 | 24 | 27 | 29 | 35 |
| 3/8 | 23 | 26 | 36 | 40 | 44 | 48 | 58 | 60 |
| 7/16 | 37 | 41 | 52 | 57 | 63 | 70 | 98 | 110 |
| 1/2 | 57 | 64 | 80 | 90 | 98 | 110 | 145 | 160 |
| 9/16 | 82 | 91 | 120 | 135 | 145 | 165 | 200 | 220 |
| 5/8 | 111 | 128 | 165 | 200 | 210 | 245 | 280 | 310 |
| 3/4 | 200 | 223 | 285 | 315 | 335 | 370 | 490 | 530 |
| 7/8 | 315 | 340 | 430 | 470 | 500 | 550 | 760 | 800 |
| 1 | 400 | 460 | 650 | 710 | 760 | 835 | 1130 | 1210 |
| 1-1/8 | 570 | 635 | 840 | 940 | 980 | 1100 | 1700 | 1750 |
| 1-1/4 | 645 | 710 | 1195 | 1315 | 1400 | 1525 | 2400 | 2650 |
| 1-3/8 | 845 | 915 | 1550 | 1780 | 1800 | 2080 | 2950 | 3050 |
| 1-1/2 | 1120 | 1195 | 2075 | 2310 | 2420 | 2700 | 3500 | 3800 |

*Previous SAE Designation

(Courtesy of Bowman Distribution, Barnes Group Inc.)

## READING ENGLISH AND METRIC MICROMETERS

(Courtesy of Brown and Sharpe Mfg. Co. Used by permission)

### *How to Read a Micrometer*

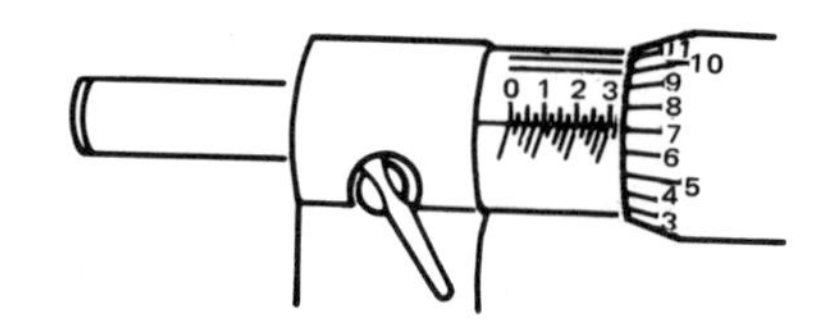


1. EXAMPLE – Reading a Conventional Micrometer to .001″

| | |
|---|---|
| Highest barrel number in sight is 3, equals | .300″ |
| Number of lines between 3 and edge of thimble is 2 (2 x .025″) equals | .050″ |
| Thimble line closest to long line on barrel (datum) is 7, equals | .007″ |
| Total Reading | .357″ |

### *The Vernier Micrometer*

1. With the addition of a vernier scale on the barrel, the thimble graduations are divided to make it possible to read in increments of one ten-thousandths of an inch (.0001″) or in two-thousandths of a millimeter (0.002 mm) on Metric models.
2. The customary method of vernier reading is accomplished after making the regular micrometer reading. Observe the number of the vernier line which coincides with a thimble line and add that number to the previous reading.

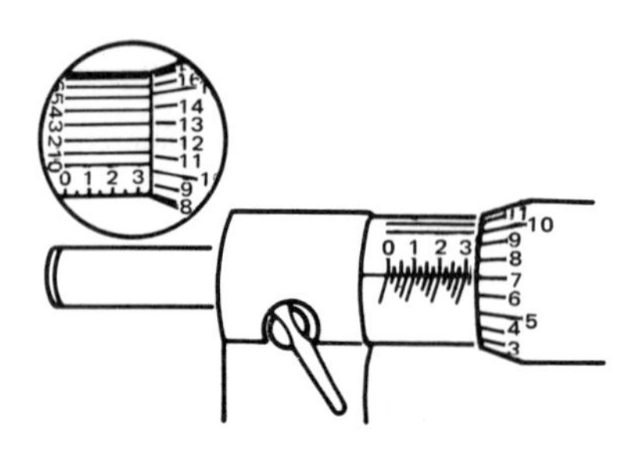


3. EXAMPLE – Reading a Micrometer to .0001″

| | |
|---|---|
| Read to thousandths, proceeding as instructed in D-1, above | .357″ |
| Number of Vernier line that coincides exactly with a thimble line is 2, equals | .0002″ |
| Total Reading | .3572″ |

## *How to Read Metric Micrometers*

1. On Metric reading micrometers, the barrel is graduated with divisions above and below the datum. Barrel graduations above the datum are "Minor Divisions" showing each 0.5 mm of movement (one full revolution of the thimble). Graduations below the datum, "Major Divisions," represent 1.0 mm each, with every fifth graduation, numbered.

2. Thimble graduations indicate 0.01 mm to 0.5 mm through each complete revolution.

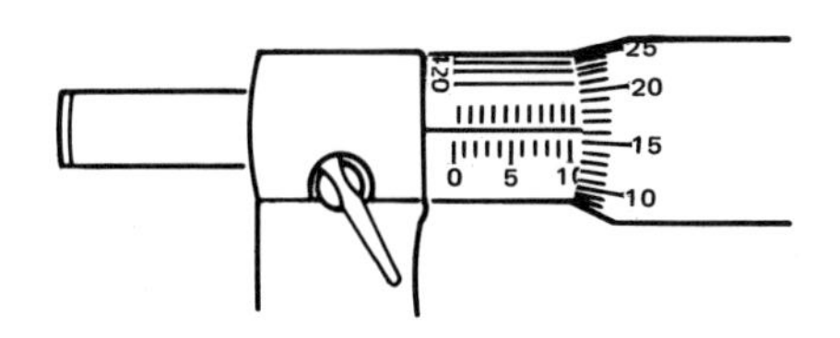


3. EXAMPLE – Reading to 0.01 mm

| | |
|---|---|
| Highest Major Division visible | = 10 x 1.00 mm = 10.00 mm |
| plus Minor Division visible | = 1 x 0.50 mm = 0.50 mm |
| plus Thimble Division nearest datum | = 16 x 0.01 mm = 0.16 mm |
| Total Reading | = 10.66 mm |

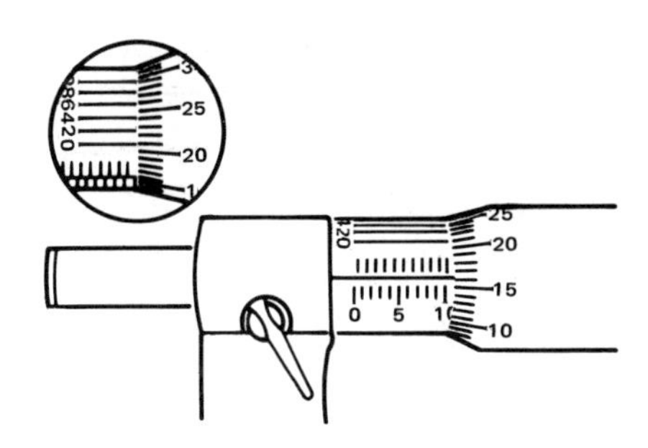


4. EXAMPLE – Reading to 0.002 mm with the Vernier

Vernier Graduations each represent 0.002 mm and are numbered 0, 2, 4, 6, 8. 0

| | |
|---|---|
| Read to hundredths of a millimeter from barrel and thimble as described above. | 10.660 mm |
| Vernier line coincident with Thimble line is 6, equals | 0.006 mm |
| Total Reading | 10.666 mm |

## READING METRIC AND ENGLISH VERNIER CALIPERS

(Brown and Sharpe TRI-CAL)

1. The caliper shown is a Brown & Sharpe TRI-CAL. It is capable of reading Inside, Outside, and Depth from the same scale, on the same side, in Metric or English.

2. EXAMPLE – Upper Vernier and Upper Beam Scale – METRIC

| BEAM SCALE | | |
|---|---|---|
| Each large numeral | 10.00 mm | (1.00 cm) |
| Each graduation (line) | 1.00 mm | |
| **VERNIER** | | |
| Aligns to indicate | .05 mm | |

Each numeral on the main scale represents 10 millimeters. The Vernier carries *20* graduations, the total span of which falls within *19* graduations on the main scale. The difference between the two scales therefore is 0.05 mm.

ILLUSTRATION SHOWS METRIC READING OF: –

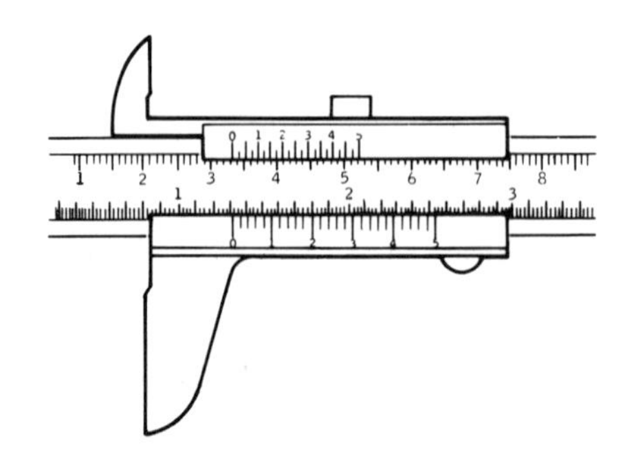

| | | |
|---|---|---|
| Upper Vernier "0" has passed large numeral "3" | 30.00 mm | (3.00 cm) |
| Upper Vernier "0" has also passed third graduation | 3.00 mm | |
| Upper Vernier "2" aligns with a beam graduation | .20 mm | |
| Total | 33.20 mm | |

3. EXAMPLE – Lower Vernier and Lower Beam Scale – ENGLISH

| BEAM SCALE | |
|---|---|
| Each large numeral | 1.000″ |
| Each small numeral | .100″ |
| Each graduation (line) | .025″ |
| **VERNIER** | |
| Aligns to indicate | .001″ |

ILLUSTRATION SHOWS ENGLISH READING OF: –

| | |
|---|---|
| Lower Vernier “0” has passed large numeral “1” | 1.000″ |
| Lower Vernier “0” has also passed small numeral “3” | .300″ |
| Lower Vernier 8th graduation aligns with a beam graduation | .008″ |
| Total | 1.308″ |

If “0” and the last graduation on the Verniers coincide with main scale graduations, the reading is an even multiple of the Verniers sub-dividing capacity and the reading can be taken opposite the Vernier “0”.

# Appendix

# B Manufacturers' Manuals


*Operating Instructions, Model AD500LR Engine Balancing Machine,* Balance Technology, Inc., 41 Enterprise Drive, Ann Arbor, Michigan 48103.

*Bear Dy-Namic Balancing Machines,* Bear Manufacturing Company, Industrial Division, Rock Island, Illinois.

*Operating Instructions for Model GR Valve Guide Redeemer, Form 93,* Hall Toledo, 2931 South Avenue, Toledo 9, Ohio.

*Repair Manual, Part 1 Crack Repair,* Irontite Products Co., Inc., 9858 El Monte Place, El Monte, California 91734.

*Circulator Manual,* Irontite Products Co., Inc., 9858 El Monte Place, El Monte, California 91734.

*Pressure Tester Manual,* Irontite Products Co., Inc., 9858 Baldwin Place, El Monte, California 91734.

*K-Line Piston Knurler Operating Instructions,* K-Line Industries, 15 West 6th Street, Holland, Michigan 49423.

*Connecting Rod Aligner Instructions, Form CRA-1 10-75,* K. O. Lee Company, P.O. Box 970, 200 South Harrison, Aberdeen, South Dakota 57401.

*Cylinder Boring Machine, Model "FW-II" Instructions and Parts List,* Kwik-Way Mfg., Co., 500 57th Street, Marion, Iowa 52302.

*Model "860" Surface Grinder, Instructions Manual and Parts List,* Kwik-Way Mfg., Co., 500 57th Street, Marion, Iowa 52302.

*Operating Instruction Rottler F2B Boring Maching,* Rottler Mfg. Co., 8029 South 200 Street, Kent, Washington 98031.

*Valve Seat Grinding Instructions, Form 194-4-76,* Sioux Tools Inc., 2901 Floyd Boulevard, Sioux City, Iowa 51102.

*The Principles of Sioux Valve and Valve Seat Reconditioning, Form No. A200-WG-3-77-20M,* Sioux Tools Inc., 2901 Floyd Boulevard, Sioux City, Iowa 51102.

*Instruction Manual, Model 85 Headmaster and Model 85-B Blockmaster,* Storm-Vulcan, Inc., 2225 Burbank Street, Dallas, Texas 75235.

*Storm-Vulcan Model 15A Crankshaft Grinder, Operating Instructions,* Storm-Vulcan, Inc., 2225 Burbank Street, Dallas, Texas 75235.

*Operating Instructions for Sunnen CK-10 Automatic Cylinder Resizing Machine, Serial Nos. 2001 and up, Form 1-CK-20B,* Sunnen Products Co., 7910 Manchester Avenue, St. Louis, Missouri 63143.

*How to Install, Operate, and Maintain Your New CH-100 Horizontal Hone, Form 1-CH-100C,* Sunnen Products Co., 7910 Manchester Avenue, St. Louis, Missouri 63143.

*Installation and Operating Instructions for the Sunnen AG-300 Precision Gage, Serial Nos. 13,000 and up, Form 1-AG-300B,* Sunnen Products Co., 7910 Manchester Avenue, St. Louis, Missouri 63143.

*P-310 Valve Guide Gage Set Operating Instructions, Form 1-P-310A,* Sunnen Products Co., 7910 Manchester Avenue, St. Louis, Missouri 63143.

*Operating and Maintenance Instructions for the Sunnen CRG750 Heavy Duty Rod and Cap Grinder, Form 1-CRG-750B,* Sunnen Products Co., 7910 Manchester Avenue, St. Louis, Missouri 63143.

*Operating and Maintenance Instructions for LBB-1810 Power Stroked Precision Honing Machine, Form 1-LBB-111,* Sunnen Products Co., 7910 Manchester Avenue, St. Louis, Missouri 63143.

*Instruction Manual, Model PM 1750 Pin Fitting Machine,* Tobin Arp Mfg. Co., 15200 West 78 Street, Eden Prairie, Minnesota 55343.

*Instruction Manual, Model 2500 TM Line Boring Machine,* Tobin Arp Mfg. Co., 15200 West 78 Street, Eden Prairie, Minnesota 55343.

*VN/IDL Valve Service Shop Instructions,* Van Norman Machine Company, Springfield, Massachusetts 01107.

*No. 530 Mini Rotary Broach, Instructions,* Van Norman Machine Company, Springfield, Massachusetts 01107.

*Operators Instruction Manual for FG5000 Flywheel Grinder,* Winona Tool Mfg., Co., 4730 West Highway 61, Winona, Minnesota 55987.

Chrysler Corporation, Service and Training Publications, Dept. 8155, 26001 Lawrence Ave., Centerline, Michigan 48015.

Fel-Pro Inc., Skokie, Illinois 60076.

Browne & Sharpe Mfg. Co., Precision Park, North Kingstown, Rhode Island 02852.

# Index